SCIENCE AND VIRTUE

Charting new territory in the interface between science and ethics, *Science and Virtue* is a study of how the scientific mentality can affect the building of character, or the attainment of virtue by the individual. Drawing on inspiration from virtue-ethics and virtue-epistemology, Caruana argues that science is not just a system of knowledge but also an important factor determining a way of life.

This book goes beyond the normal strategy evident in the science-ethics realm of examining specific ethical dilemmas posed by scientific innovations. Here Caruana deals with more fundamental issues, uncovering morally significant tendencies within the very core of the scientific mentality and explaining how science, its method, history and explanatory power can shape a conception of the good life.

ASHGATE NEW CRITICAL THINKING IN PHILOSOPHY

The *Ashgate New Critical Thinking in Philosophy* series brings high quality research monograph publishing into focus for authors, the international library market, and student, academic and research readers. Headed by an international editorial advisory board of acclaimed scholars from across the philosophical spectrum, this monograph series presents cutting-edge research from established as well as exciting new authors in the field. Spanning the breadth of philosophy and related disciplinary and interdisciplinary perspectives *Ashgate New Critical Thinking in Philosophy* takes contemporary philosophical research into new directions and debate.

Science and Virtue

An Essay on the Impact of the Scientific Mentality on Moral Character

LOUIS CARUANA
Heythrop College, University of London, UK

ASHGATE

Published by
Ashgate Publishing Limited
Gower House
Croft Road
Aldershot
Hampshire GU11 3HR
England

Ashgate Publishing Company
Suite 420
101 Cherry Street
Burlington, VT 05401-4405
USA

Ashgate website: http://www.ashgate.com

British Library Cataloguing in Publication Data
Caruana, Louis
Science and virtue : an essay on the impact of the scientific mentality on moral character. – (Ashgate new critical thinking in philosophy)
1.Virtue 2.Science – Moral and ethical aspects
I.Title
179.9

Library of Congress Cataloging-in-Publication Data
Caruana, Louis, 1957-
Science and virtue: an essay on the impact of the scientific mentality on moral character / Louis Caruana.
p. cm. – (Ashgate new critical thinking in philosophy)
Includes bibliographical references and index.
ISBN 0-7546-5562-8 (hardcover : alk. paper)
1. Science – Moral and ethical aspects. 2. Virtue. I. Title. II. Series.

BJ57.C35 2006
171'.7 – dc22

2005030620

ISBN-10: 0-7546-5562-8
ISBN-13: 978-0-7546-5562-6

Printed and bound by TJ International Ltd, Padstow, Cornwall.

Contents

Preface

In this book, I engage in a study of the scientific mentality. The idea of writing it came as a reaction to a certain polarisation that is evident in post-enlightenment culture. On the one hand, natural science comes across as beneficial at a deep level because humanity can today find many answers in science which earlier generations could seek only in the realm of religion. Thus, religion and culture have been freed from contingent complications and, as a result, purified and deepened. On the other hand, it is clear that the structural injustice in today's world is rooted in value-systems promoted by powerful social and political forces that exploit science and technology for their own advantage. Science, in other words, lets itself become an instrument of injustice. This polarisation has unfortunately received very little attention from scientists and philosophers. As things stand now in academia, while scientists seek explanations of various phenomena and analytic philosophers of science help them refine their methods and draw some implications from their world-view, there is a lack of interest in the personal values that govern the life of the individual scientist and of others affected by the scientific mentality. Given that natural science has become one of the major players in contemporary culture, it is a mistake to assume that its impact on how people live is negligible.

This is the original insight that directed the inquiry contained in this book. To avoid getting lost in the complexity of the subject, one particular approach will be adopted through out: an approach that is analytical rather than psychological, sociological or phenomenological. The line of argument is situated, in a sense, on two levels. On one level, it highlights the link between knowledge and practice in the sciences. On the other, it branches out, at various steps, into some extrapolated situations. These allow an exploration of the possible effects on moral character. Some traces of this approach can be seen in Karl Popper's views on the moral responsibility of the scientist. In a paper published in 1970, he makes an interesting appeal to the famous Hippocratic Oath.[1] This oath consists in three main parts. First, the apprentice declares his personal obligation to his teacher. He then promises to carry on the tradition of his art, preserve its high standards as regards defending the sanctity of life, and to pass on the art responsibly to his students. Thirdly, the apprentice pledges to be single-minded in his aim to help the suffering, and to retain confidentiality. Popper makes a quick attempt to apply this to natural science in general. His application to modern times results in three maxims. He places professional responsibility first. The student is invited to acknowledge and accept his duty to further the growth of knowledge by participating in the search for truth, even though

1 Karl Popper, 'Moral Responsibility of the Scientist', in P. Weingartner and G. Zecha (eds), *Induction, Physics and Ethics* (Dordrecth: Reidel, 1970), pp. 329–36. See Appendix A for a full version of the Hippocratic Oath.

he accepts his fallibility. Second, the student should express his respect towards the community and tradition he belongs to but should also expressly undertake the duty of being critical towards himself and others: '*most important*, he should beware of intellectual arrogance, and try *not to succumb to intellectual fashions* [italics in the original]'. Thirdly, he should acknowledge his overriding loyalty not to his teacher or colleagues but to humanity in general. In this reformulation of the Hippocratic Oath, what stands out is Popper's allusion to some basic virtues and vices. The final point is especially important. It subordinates the intellectual and practical life of the scientist to the overriding loyalty to humankind. Hippocrates had insisted on the physician's loyalty towards patients. Popper insists on the scientist's loyalty towards humankind. The basic approach in both cases is not very different from Aristotle's concern with genuine human flourishing as a point of reference in defining virtue.

The inquiry in the following chapters is situated within this broad philosophical trend. The primary objective is not to deal directly with the various specific ethical dilemmas posed by some frontier areas of science and technology. My aim is rather to deal with more fundamental issues. It is essentially to uncover morally significant tendencies within a scientific age. I have no pretension of telling scientists what to do to be more moral. That is a question that involves personal choice and commitment. It is up to each individual to decide how to live. My aim is only to indicate what aspect of their work can affect moral character.

I wrote this book over a period of several years, and I cannot hope to recall all the colleagues and students with whom, at various times, I have discussed some of the topics it deals with. For comments on previous drafts, I have benefited especially from those of Linda Zagzebski, Robert Audi, Garth Hallett, Peter Lipton, and Bas and Tina Jongeling. The finished result represents some first steps towards a full appreciation of the interface between science and ethics, and towards an adequate understanding of the consequences of the scientific mentality. I hope that my readers will find interest and profit in this exploration.

Chapter 1

A Preamble on Virtue

Much work in moral theology and ethics has recognised the important fact that the way individuals live, the way they confront dilemmas and determine their obligations depends on how they see the world. In the course of history, religious and philosophical beliefs have been considered very important for the determination of a helpful list of virtues as road signs for the education of youth. Some recent historians, however, have argued convincingly that the present cultural situation in the world is so lacking in consensus that any attempt to revitalise the idea of a virtuous life is doomed to failure.[1] What seems a virtue to one can sometimes appear a vice to another. What could be the use of an appeal for a more virtuous life, if such an appeal will be refracted in myriad ways within the labyrinth of a fragmented culture? Admittedly, there is more disagreement on whether a piece of conduct is virtuous than disagreement on whether a given virtue is desirable. Radical relativism, however, remains a serious threat. The only hope seems to lie in a general retreat. One gives up the idea of reaching the global multitudes. One concentrates rather on small social units or communities within which intellectual and moral life could be rediscovered and sustained intact.[2]

The time has come, however, to realise that the fragmentation of culture that seems to necessitate such a retreat is just one layer among others in today's complex cultural landscape. There is another layer that presents itself as uniform and global. It is constituted by the methods and results of the natural sciences and technology. No amount of pluralism among philosophers of science, no degree of radical relativism among sociologists of knowledge, seems capable of undermining this simple fact. Natural science and technology are here to stay, and they are everywhere. Determining how such a cultural reality, how such a kind of uniform substratum, reflects upon the building of character, upon the attainment of virtue, promises therefore to be a fruitful path to follow in view of rediscovering the common aspirations of the global community as regards the good life. Such a move is based on the observation that religious and philosophical beliefs are not the only ones that matter. The personal dispositions of people working in the area of the natural sciences are deeply marked by the methods and skills required by this discipline, and also perhaps by the discoveries themselves regarding the nature of the world. It is therefore very reasonable to assume that the way many individuals live depends on natural science

---

1 Most notably, Alisdair MacIntyre in his *After Virtue: a Study in Moral Theory*, (Notre Dame: University of Notre Dame Press, 1981).

2 Ibid., p. 245.

and on whatever description of the world comes to be considered warranted by its methods. The domain of philosophy of virtue is vast. One needs to clarify at the outset some fundamental aspects that are indispensable for this kind of investigation. This introductory chapter will be focused on the following three questions. What is virtue? How does one grow in virtue? How can the natural sciences be relevant for growth in virtue?

Understanding Virtue

Recent philosophical work in the area of ethics has often focused primarily on duties, obligations, moral dilemmas, borderline cases and hypothetical situations. Such themes are certainly important and deserve a certain amount of attention. Ordinary people, however, are rarely in a position where they feel surrounded by tragic borderline cases, perplexing moral dilemmas or major clashes of duties. Philosophers who study virtue seek to concentrate not only on extreme moments or situations that stretch ethical theories to their limits, but also on the continuous behaviour patterns and dispositions that characterise an individual, that characterise a life. The study of virtue involves the study of such dispositions or character-traits. It therefore constitutes a specific domain distinct from other areas of interest in ethics, a domain that merits close attention, especially by those whose interest lies primarily in persons themselves.

Philosophers since antiquity have shown that the idea of virtue can be understood in various ways. Much insight can be gained by recalling at least the two distinct ways presented by Plato and Aristotle. The former struggles at some length with the relation between virtue and knowledge. His overall view seems to amount to the idea that the nature of virtue is essentially expressive of a kind of knowledge. A virtuous person is one who lives in conformity to certain rational principles reflecting his or her true interests. On the contrary, a person of vice lives under the influence of various destructive natural passions and appetites. The basic proposal, therefore, is that virtue is the rule of natural inclination by right reason, while vice is the ignorance of the true good that leads to the individual knowing no better than to act at the impulse of his or her irrational instinct and appetites. Aristotle differs from this on a number of points. His basic originality lies in the idea that virtue is not a kind of knowledge but a character-trait of the individual. His interest lies primarily in determining a set of personal attributes in terms of which human beings might be said to fulfil their proper function or realise their proper end *qua* human beings. Just as we learn what a good horse is by observing and comparing actual horses in respect of the tasks we require of them, so learning about the virtue of humans is similarly a matter of observation and comparison. In this, Aristotle differs radically from Plato, who mistrusted observation of actual states of affairs arguing that what needs to be done is to avoid being tied down to what we perceive so as to apprehend the abstract form of the good, in a somewhat mathematical, intuitive grasp. Moreover, while for Plato human inclinations, passions and appetites are to be considered bad in themselves, for Aristotle they aren't. Everything depends on the way in which these inclinations are expressed or exercised.

There are certainly other significant aspects worthy of fruitful investigation as regards the difference between these two basic views on the nature of virtue, but these need not concern us here. The little that has been said is enough to show that some fundamental options arise at the very start of any inquiry concerning virtue. The view I endorse in this book is more in line with the Aristotelian view than with the Platonic one. My approach will be naturalistic, in the sense that I will be engaged in extracting some valuable insights not from an allegedly necessary idea of the good, but from what good people actually do when they are affected by the scientific mentality. The strategy therefore is more inductive than deductive. There will certainly be an element of contingency in this inquiry precisely because it is enmeshed in a world of changing particular situations rather than well-framed by absolute, changeless principles.

Having clarified this fundamental option as regards my basic method, let me advance a somewhat fuller account of virtue within the parameters decided upon. If our inquiry starts from what good people typically do, then the focus of our attention is on character-traits or dispositions of individuals. When the disposition is good, we call it a virtue.[3] We recognise an individual as virtuous in some way when he or she acts in the way characteristic of that virtue. This recognition is not straightforward, because we may have an individual doing what virtue requires but for a reason inappropriate to that virtue. We may have an individual, say, who does the right thing from self-interest. This complicates matters. To avoid pitfalls, it is good to recall that acting from virtue is a matter of explaining the act by reference to the virtuous disposition of the agent. Such action is the kind of action that manifests the virtue of the individual.[4] It is worth noting also that the capacity itself to act in such or such a manner is not the virtue. It is the capacity's being exercised well that constitutes the virtue. It is not, for instance, the faculty of moving one's body around that constitutes courage, but the excellence of this faculty in certain circumstances, namely moving in the right way, say, in risky situations confronted by rescue teams. There cannot be any hard and fast rule to determine in all cases when rescuers should move into a dangerous zone or should refrain from doing so. The need for speaking of virtues

3 The definition of virtue that has served as the major starting point for much philosophical work through many centuries is that of Peter Lombard: 'virtus est bona qualitas mentis, qua recte vivitur, qua nullus male utitur, quam Deus in nobis sine nobis operatur [virtues are good qualities of mind which dispense us to live rightly, which we cannot misuse, and which God works in us without our help]'. See Petri Lombardi, *Libri IV Sententiarum* (Florence: Ad Clares Aquas typographia Collegii S. Bonaventura, 1916), Book II, chapter 27, section 5. It is interesting to note that St Thomas Aquinas makes the remark that the term 'quality' in this definition can be substituted by 'disposition' so as to give a more precise genus, and hence a better definition. See Thomas Aquinas, *Summa Theologiae*, 1a2æ, Q55, art 4 (London: Blackfriars, 1964).

4 For a detailed analysis of how an act can be conceived as grounded in virtue, see R. Audi, 'Acting from Virtue', *Mind*, 104 (1995): 449–71.

arises precisely from the simple fact that life is complex. One can certainly concede the point that some guiding principles can help. When one is on the spot, however, it is up to the individual to exercise his or her faculties, and to do it well in those particular and unrepeatable circumstances.

The basic point is that one may distinguish between talking about good people and talking about good acts. It is the same individual who acts in the multiplicity of particular, unrepeatable situations. Taking virtue as a good character-trait of the individual is therefore a good starting point, since it expresses the element of continuity over and above the variety of act-situations. It is certainly possible to delve deeper into the nature of virtue itself. One may refer to the distinction between the various virtues, known in various philosophical works as the three theological virtues, faith, hope and charity, together with the four cardinal virtues, prudence, temperance, justice and courage. A study of these distinctions, however, will not be carried out in any detail here since it would lead the discussion too far away from our main investigation envisaged in this book.[5] It is essential however to say something further about virtues in so far as they are attributes.

If virtue is a kind of habit, then it is definable within human life *taken as extended in time*, sometimes taken even from beginning to end. If, moreover, virtue enables us to achieve a certain good, it is related to a certain aim. It is clear therefore that, when doing ethics in terms of virtues, the bearer of these virtues, the person, is seen as the one subject of a narrative that runs from birth to death, of a unique and unrepeatable life-story.[6] A life includes many different aims; so much so that the individual may sometimes feel frustrated by various incompatible aims. A scientist, for instance, cannot be considered an individual characterised solely by a life in the laboratory. There is more to life than what happens in the laboratory, even for fully dedicated scientists. There are other areas of life, other aims that are pursued. Taking the scientist to be an individual in the laboratory does not do justice to the fact that the one living in the laboratory and the one having experiences outside the laboratory are one and the same person. Achieving a certain degree of integrity in one's life means organising one's various aims in such a way that they reinforce each other. Short-term aims are pursued in function of long term ones. The project in the following chapters can be described as a move to rediscover the unity of life of those engaged in science. This will be done by determining how virtue is affected, for better or for worse, by some aspects of the world of natural science. The main assumption therefore is that some aspects of scientific practice are deep enough to

5 For useful study of these virtues see Peter Geach, *The Virtues* (Cambridge: Cambridge University Press, 1977).

6 For this interesting link between narrative and the nature of virtue, which certainly deserves more attention but lies beyond the scope of this chapter, I am indebted to MacIntyre, chapter 15. See also T. Söderqvist, 'Virtue-Ethics and the Historiography of Science', *Danish Yearbook of Philosophy*, 32 (1997): 45–64 for a discussion on how scientific biography can be a narrative that is essentially *edifying*.

affect the quality of the person engaged in this practice. They are deep enough to leave their mark on all areas of the life-story of the individual.

Growth in Virtue

In the *Meno*, Plato presents Socrates answering the question of how best to acquire virtue. Surprisingly, the study of dialectic does not come first. In order to help people become virtuous, Plato starts with mathematics. It is only through such a discipline, which determines everything with certainty and which lacks space for personal opinion, can a student avoid losing the passion for truth. If beginners are faced with plausible argument and plausible counter-argument, they often conclude that in such matters there is only opinion and no knowledge. The first step therefore is to consolidate in the beginner a certain passion for truth. Personal sentiments and opinions must be seen at first as of marginal importance. One must give oneself over to the facts, receiving and discovering rather than producing or imposing one's own views. Such an exercise helps the individual grow immune to vanity, self-deception and disinterestedness. It is a long-term process. Only when the basic attitudes towards truth and falsity are well established can the individual move on to more demanding environments.[7]

For such environments, moral philosophers have identified a process with at least three distinguishable stages. The first consists in the learner's living close to a master from whom he or she can absorb, as it were, the virtue manifested in the master's way of acting. A set of rules for action could be a help to the learner, but rules will never be enough to make someone virtuous. The learner needs to know how to apply the rules and when. Only the master can help the learner gain those habits that go beyond ratiocination. The second stage of the process of becoming virtuous consists in the learner's interior struggle to overcome fear, doubts, and weakness of will when it comes to action. Suppose a young cadet has been accompanying a courageous captain in a number of campaigns. For the cadet to grow in courage, accompanying

7 Plato describes this doctrine most clearly in the second half of chapter VII of the *Republic*, where arithmetic and geometry are considered the first two subjects to be taught. Through these subjects, the students imbibe a craving for 'what always is, not what comes into being and passes away' (527[b]). As they grow into mature thinkers, they will engage in difficult subjects without losing sight of the major attitude of seeking truth. They will thus avoid the temptation of taking philosophy to be nothing more than 'dragging and tearing those around them with their arguments' (539[b]). Iris Murdoch, in her *The Sovereignty of Good* (London: Ark, 1970), highlights this point. She recalls that, for Plato, the main road towards wisdom is 'the way of the τέχναι, the sciences, crafts, and intellectual disciplines excluding the arts.' She continues by highlighting the importance of the unity of the person: 'I think there is a way of the intellect, a sense in which intellectual disciplines are moral disciplines, and this is not too difficult to discern. There are important bridge ideas between morality and other at first sight different human activities [...] The τέχνη which Plato thought was most important was mathematics' (pp. 88–89). See also: Gilbert Meilaender, *The Theory and Practice of Virtue* (Notre Dame Indiana: University of Notre Dame Press, 1984).

the captain is necessary but not sufficient. It is necessary as a start, but there will be a time when the cadet needs to be on his own, and that is precisely when doubts and fears need to be overcome. The third and final stage in the process of growing in virtue consists in the learner gaining self-confidence and a certain self-satisfaction in acting virtuously. At this stage, doubts and fears are definitely overcome. The learner is independent. What is happening during this process has been famously described by Aristotle as the attainment of the mean between two extremes. The mean refers to the virtue, in this case courage, and the two extremes refer to two opposite bad habits, in this case an 'excess' of courage, which is rashness, on the one hand, and a 'deficiency' of courage, which is cowardice, on the other. In Aristotle's words, 'virtue, then, is a state of character concerned with choice, lying in a mean, i.e. the mean relative to us, this being determined by a rational principle, and by that principle by which the man of practical wisdom would determine it.'[8]

Science and Virtue

This way of understanding growth in virtue allows a valuable exploration of the possible effects a life in science may have on the individual. To become scientists, young people learn specific rules, rub shoulders with experts, have to overcome doubts and hesitation, eventually arrive at self-confidence. Something important is happening during this long process as regards the individual. The transformation of the individual has received little attention in the vast literature on philosophy of science. The aim of this book consists in determining how the particular mind-set characteristic of natural science stimulates growth in virtue on certain fronts. The basic presupposition is that the methods of scientific investigation impinge on the individual scientist's way of life not only as regards knowledge acquisition but also at a deep personal level. Some dispositions are reinforced, others subdued or discouraged. The methods of science affect the habits of the scientist. The challenge to be faced in the following chapters is precisely the exploration of the terrain opened up by this assumption so as to see where it can lead.[9]

---

8 Aristotle, *Nicomachean Ethics*, 1107^{a}, 1–3, trans. W.D. Ross, in R. McKeon (ed.), *The Basic Works of Aristotle* (New York: Random House, 1941), p. 959.

9 The link between the scientific mentality and personal responsibility has been the subject of a number of recent studies, for example Andrew G. van Melsen, *Science and Responsibility* (Pittsburgh: Duquesne University Press, 1970); Richard Bernstein, *Beyond Objectivism and Relativism* (Oxford: Basil Blackwell, 1983); Nicholas Maxwell, *From Knowledge to Wisdom* (Oxford: Basil Blackwell, 1984). In all these works, however, virtue as such has not been an area of explicit concern. Some premonition of the importance of virtues in relation to science can be seen however in the work of J. Dupré, *The Disorder of Things: Metaphysical Foundations of the Disunity of Science* (Cambridge, Mass.; London: Harvard University Press, 1993), where he states: 'I suggest that we try to replace the kind of epistemology that unites pure descriptivism and scientific apologetics with something more like a virtue epistemology' (p. 243).

Before embarking on this journey, however, a crucial question needs to be met. Is the good scientist necessarily a good person? This question needs careful handling because there seems to be an important distinction between goodness at the level of skill and goodness at the level of a person as a whole, as is intended by the term virtue. Aristotle has famously expressed this point by saying: 'it is from the same causes and by the same means that every virtue is both produced and destroyed, and similarly every art; for it is from playing the lyre that both good and bad lyre-players are produced'.[10] For the purposes of this book, this statement can be translated as follows: it is from engaging in scientific practice that both good and bad scientists are produced. There is no automatic transformation. The engaging in a practice does not guarantee goodness at the level of skill. Still less does it guarantee it at the level of virtue.

One way of replying to the question whether a good scientist is necessarily a good person is to examine the distinction between skills and virtues. Let us call an acquired skill a talent, whether it was acquired through strenuous practice or just as a gift at birth. Talents enable people to pull off certain distinct tasks. Virtues, on the contrary, are not traits that are primarily directed to the achievement of certain tasks, but are directed primarily to the standing in a proper relation to what is good. In other words, they enable persons to give certain values a proper governing place in the overall direction of their lives. Explained in this way, talents and virtues are seen as standing at opposite ends of the spectrum of the individual's life. The impression we get is that no significant interaction can occur between talents and virtues. This impression, however, is quite misleading. It is not difficult to see how some talents can affect the person's growth in virtue, and also vice versa. As a good illustration of this, consider memory as a talent, as explained in N.J.H. Dent's study, *The Moral Psychology of the Virtues*:

> [A]lthough we quite properly admire a man's excellent memory, we need have no anxiety that such-like qualities cannot be distinguished from his virtues and vices. It is perhaps worth remarking that what may be put down to a 'poor memory' can quite often be a symptom of vice. Someone may be so wrapped up in himself that he simply fails properly to attend to others and pay sufficient attention for the necessary matters to 'register' with him so that they might later be recalled. Or he may to be too lazy to think hard enough about what he has been told to do. Conversely, what may be put down to someone's having had merely the good fortune to be blessed with an excellent memory may in fact be a matter of his being sufficiently mindful of and attentive of others that he clearly and properly registers and understands the needful points.[11]

---

10 Aristotle, *Nicomachean Ethics*, 1103^b, 7.

11 N.J.H. Dent, *The Moral Psychology of the Virtues* (Cambridge: Cambridge University Press, 1984). On growth in virtue, a very useful overview is: David Carr, *Educating the Virtues: An Essay on the Philosophical Psychology of Moral Development and Education* (London: Routledge, 1990).

The main point here is that, even when we accept that a distinction does exist between talents and virtues, we can still trace a causal link between talent and virtue.

More plausibility to this reply will be evident if we recall the fundamental unifying element inherent in the very idea of person. As we know, false dichotomies and divisions may be counterproductive in a philosophical inquiry. We need to use analysis. There is certainly much benefit in this method. It may even be indispensable. Nevertheless, we need to use it carefully lest we miss the wood for the trees. For instance, suppose I ask: What is the proper work of a scientist? The *prima faciae* answer would be: To become a good scientist. This is certainly a legitimate answer. Aristotle's observation was: 'The virtue of a thing is relative to its proper work'.[12] There is no doubt that, just as young lyre-players play the lyre to become good lyre-players, so also young scientists should engage in science to become good scientists. The life of the individual, however, is broader than playing the lyre, broader than science. It is one and the same individual who lives as a lyre-player, or as a scientist, and also as a person in many other areas of life. There are no hermetic compartments within the person. This suggests that the most desirable situation is one where the individual has a unifying aim that goes beyond specific practices, and embraces all of them. We refer to this one purpose when we acknowledge that the ultimate aim of an individual should be not just to become a good lyre-player, or a good scientist, but ultimately to become a good person.[13]

As an illustration of this point, we may refer to Plato's insistence on the unity of the virtues. The basic point, acknowledged also by Aristotle and many others since then, is that there is a lack of sharp boundaries between the 'moral' and 'non-moral' spheres. Psychological connections between the ethical and other aspects of character abound. Admittedly, Aristotle made an attempt at fixing some boundaries in his vocabulary. He famously distinguishes between the moral from the intellectual virtues. The former are character-traits that make for the attainment of our moral aim, namely to do good, and ultimately to become good persons. The latter are those attributes that make for the attainment of our cognitive aims, namely to grasp the truth and ultimately become truthful persons. One of the intellectual virtues, however, namely what he calls *phronesis*, is itself necessary for excellency of character in moral sense. He is therefore aware of the need for highlighting various psychological connections that express the unity of the person. Recent scholarship in the area of virtue ethics is unfortunately somewhat limited because of the bad effects of compartmentalisation in contemporary academic philosophy. Issues dealing with virtues are discussed in moral philosophy or theology in a way that detaches them

12 Aristotle, *Nicomachean Ethics*, 1139[a], 17.

13 This argument can be extended beyond the cases of lyre-players or scientists. One may indeed apply the argument for all kinds of knowledge. Knowing in general can be considered an act. It is an act in which the knower gets credit for achieving truth. Moreover, truths that are believed by a knower can be of various kinds. One gets more credit if the truth believed is desirable. Hence, we can see how knowledge in general is not independent of moral value, because believing truths of certain kinds is linked to a good life. See: Linda Zagzebski, 'The Search for the Source of Epistemic Good', *Metaphilosophy*, 34 (2003): 12–28.

from the theoretical contexts on which they were originally constructed. There is a tendency to forget that previous masters in this area, such as Aristotle and Thomas Aquinas, defended theses about the virtues as integral parts of complex, unified bodies of psychological and metaphysical theory. Even if convenient, it would be a mistake to divide the human individual into separate areas of inquiry that may easily become incommensuarble.

Insisting on the unity of the person may not be totally convincing to everyone. In fact, a major objection needs to be tackled as regards the apparent distance between problems of a scientific nature and those of an ethical nature. This objection will return in various forms all along the following chapters. Skill in solving differential equations, say, is so different from skill in living well that some very strong justification needs to be made to convince anyone that the mutual interaction between these two human activities is significant. Precisely because of this distance, it has been customary to distinguish between philosophic and practical wisdom, the former consisting in scientific knowledge combined with intuitive reason about things that are highest by nature, the latter consisting in the deliberation of things human. This distance between the two realms is significant, and has been the subject of discussion since ancient times. Anaxagoras and Thales may be taken as typical precursors of many of our contemporaries who engage in science and yet ignore how to live well. Aristotle recalls: ‘This is why we say Anaxagoras, Thales and men like them have philosophic but not practical wisdom, when we see them ignorant of what is to their own advantage and when we say that they know things that are remarkable, admirable, difficult and divine but useless; viz. because it is not human goods that they seek.’[14]

One may include in the objection a further dimension: an argument concerning age and experience. Young people can be ‘wise’ in matters like mathematics and geometry, where certainty is easily accessible. They cannot however be wise in the sense of practical wisdom, because this needs experience. Using the vocabulary of classical philosophy, one can further explain this by saying that the former kind of wisdom, that of mathematics and geometry, is concerned with universals, while the latter, that of practical wisdom, is concerned primarily with particular actions that become familiar from experience.[15] Scientific knowledge, together with the practice that supports its achievement, is geared towards the discovery of laws describing regularities. It is geared towards the formulation of the limiting premises, the most general principles. On the contrary, practical wisdom is concerned with a particular, unrepeatable action, the ultimate particular. This objection, therefore, tends to undermine the entire project of this book because it seems to indicate that the two realms are like vectors pointing in opposite directions, completely distinct and even perhaps, in a sense, incommensurable.

This argument, however, is not as damaging as it looks. One needs to unveil the mutual interaction that can occur between ratiocination, intuitive reason and

14 Aristotle, *Nicomachean Ethics*, 1141^{b}, 4–8.

15 Ibid., 1142^{a}, 13–30.

practical wisdom. A good starting point is the distinction between the nature of knowledge and the nature of virtue. Knowledge is gained in a discontinuous manner. It is acquired by a person one bit at a time. It involves piece-meal acquisition of truth. Virtue, on the contrary, is a certain disposition and consists in how a person is. What is piece-meal in this case is the *manifestation* of the virtue, but not the virtue itself. The virtue, just as all other dispositions, is best seen as a state of the person. A person's dispositions are affected by what the person does in a piece-meal fashion, but the effect will only be observable in the long run. Becoming good lyre-players needs the repetition of specific movements of the hands, over and over again, until these movements seep in and become part of the person's way of life. In general, becoming good persons needs repetition of specific good actions, until goodness seeps in and becomes an attribute of the person.

The fundamental point here is that it is certainly possible to focus our attention on the person as a whole, spread over time, rather than on a single time-slice of the person. This stance, or point of view, enables us to talk about, and even determine, certain features of the personality that become evident only through repeated use of a particular method in thinking and acting. Some habits will be reinforced, others weakened. There is therefore a link between ways of thinking and the overall state of a person. The classical way of putting this is: one becomes just by doing just acts. The adjective we use to describe an individual act catches on, as it were, to the person doing the act. Nevertheless, the attribute of the act becomes also the attribute of the person only after repeated instances of that kind of act. What I am proposing here is similar to this classical way of discussing virtue. Some aspect of the nature of a particular act, even if it is one performed in the laboratory, leaves its mark on the deeper dispositions of the person doing that act, in this case the scientist. The mark will become evident if the act is performed repeatedly. In other words, living a life of science produces a certain mind-set associated with certain habits. Some of these are good, and are called virtues. Others are bad: vices.[16]

One needs to keep in mind that a particular act done in a limited period of time, an act observable in a thin time-slice of the person, cannot help us to determine which aspects of the state of the person are being reinforced and which subdued. One just act does not necessarily mean that the person who did it is just. There is one

16 For important discussions on the nature of virtues see Georg Henrick von Wright, *The Varieties of Goodness* (New York: Routledge & Kegan Paul, 1963), chapter 7; Lester Hunt, 'Character and Thought', *American Philosophical Quarterly*, 15 (1978): 177–86; J. McDowell, 'Virtue and Reason', *The Monist*, 62 (1979): 331-50, reprinted in R. Crisp and M. Slote (eds), *Virtue Ethics* (Oxford University Press, 1998), pp. 141–62. Von Wright and Hunt introduce various distinctions between character-traits, habits, temperamental dispositions and skills. They insist that virtues fall under character-traits, not under any of the other categories. Since their arguments deal exclusively with the nature of virtue, their analysis lies outside the scope of this book. I work with the assumption that, since character traits are achieved by habituation, as von Wright himself admits (p. 143), there are good reasons to consider the genus habit broad enough to include character-traits, these latter being habits that are well-established and that involve the harmonisation, in action, of both reason and passion.

question that concerns the rightness or wrongness of a single act, and there is quite another question that concerns the person. To deal with the second question, one needs to go beyond the single act. One needs to handle the situation in terms of the praiseworthiness or blameworthiness of the person.[17] When dealing with science and virtue, therefore, one needs to go beyond the specific scientific inquiry itself. One needs to uncover the most general features of scientific practice so as to determine any long-term effects it might have on the person.

A Research Strategy

The study of the philosophical foundations of the empirical sciences is essential for some grasp on what we mean by the term 'scientific mentality'. The presentation in this book will be divided into four major topics, dealing respectively with observation, the logical structure of scientific method, explanation, and the evaluation of scientific theories in the course of history. In each of these chapters, an attempt will be made to highlight the link between the theoretical and the practical. Such a link will allow me to include in each chapter a concluding section dedicated to the question of how some of the major aspects discussed can impinge on the deeper level of the person, and leave a mark on his or her disposition towards the good. It can be seen, therefore, that the drawing of some conclusions as regards virtue will be carried out systematically, one topic after another. Each topic will be analysed in order to determine the vices of excess and deficiency that can be associated with that particular aspect. Once these vices are evident, the virtue associated with that particular area will be determined in the Aristotelian way of seeking the mean. In the final chapter, I will attempt a synthesis of the conclusions drawn from each previous chapter.

A final comment needs to be made about the results one should expect from this investigation. Put simply, one needs to keep in mind that any determination of possible effects of science on virtue is necessarily imprecise, inexact. The best one can hope for is more of the nature of a proposal than of the nature of a deduction that obliges with the force of logical necessity. This may disturb those readers who tend to show an exclusive preference, always and everywhere, to scrupulous precision in reasoning. One significant assumption I will be working with in the following chapters is that, in a given philosophical study, the lack of precision does not necessarily invalidate the conclusions. This is justified to the extent that, even though precision is an important value, limiting philosophy to rigorous precision

---

17 In general, an ethical theory can be either act-based or virtue-based. These two modes of ethics correspond to the two questions described here. The latter kind of ethics has recently seen an important revival in what has come to be called virtue ethics and virtue epistemology. These positions are well represented respectively by Rosalind Hursthouse, *On Virtue Ethics* (Oxford University Press, 1999), and Linda Zagzebski, *Virtues of the Mind* (Cambridge: Cambridge University Press, 1996). A move towards the application of virtue-ethics to historiography can be found in Söderqvist, 'Virtue Ethics and the Historiography of Science'.

will leave out vast unexplored areas. It will leave out the entire landscape indicated by very general, philosophical questions that cannot be handled by strict logic. In evaluating philosophical work in general, being excessively bound to the need for precision may even sometimes be counter-productive. Aristotle's remark about the educated man is still valid: 'it is the mark of an educated man to look for precision in each class of things just as far as the nature of the subject admits; it is evidently equally foolish to accept probable reasoning from a mathematician and to demand from a rhetorician scientific proof'.[18] Ethical reality is immensely difficult to see clearly, and it therefore requires an attitude of humility. The attainment of precision or the lack of it, therefore, will not be considered of prime relevance for knowing whether my inquiry has been successful or not. The entire exercise will be considered successful to the extent that it allows the compilation of a list of reasonably well defined virtues characteristically associated with a certain way of life, that of the scientist. One should not forget that a person might also grow in virtue because of habits associated with other ways of life. The cultural situation of modern life, fragmented as it is, permits in fact a multifaceted way of life in which an individual may be switching from one world to another several times during a single day: the laboratory, the lecture room, the home, the pub, the church, and so on. What this book is about is the impact of only one of these worlds.

---

18 Aristotle, *Nicomachean Ethics*, 1094^b, 24. See also J. McDowell, 'Virtue and Reason'. A first step towards a more precise rendering of the Aristotelian method adopted here can be found in Appendix B.

Chapter 2
Observation

Observation is a skill. Some people are better at it than others. There are many possible reasons for this. Some people are born with a certain talent of being good observers. Others acquire this skill after long years of training. It is reasonable to assume that both nature and nurture are involved here. The training of an individual for a scientific career depends both on some degree of personal talent, and also on the development of some specific skills in the course of training. The practice of science enhances what lies there as potential at birth, and observation is a constitutive ingredient of this scientific practice. If someone protests that theoretical departments of science do not engage directly in observation, it is enough to recall how the link between theory and experiment can be considered the cradle of the scientific revolution. Essentially, a theory is evaluated with respect to how the world is. Observation enters here, at this evaluation stage. The nature of observation has been discussed and analysed in various ways. My main interest is to arrive at some important insights not about the mechanism of observation as such, if there is one, but about the way observation affects the set of common habits that characterise the majority of scientists, which is what I'm calling the scientific mentality. From here I will draw some conclusions regarding the effects this mentality may have on the individual's capacity to do good.

There are two major steps in the argument. The first one is meant to show that we cannot give a genuine account of observation if we neglect the observer's life. In this context, I take the term 'life' to refer not only to the background *beliefs* of the observer. That would make my thesis directly equivalent to one of the important claims in philosophy of science of the post-positivist period. With background beliefs, I would like to include also two other factors. Firstly, I'm taking life to include the various kinds of *habits* that give rise to these beliefs. Secondly, I'm taking it to include also the particular *actions* of the observer that manifest these habits and that, in the long run, may enhance or impede the development of some of these habits. The sense of 'life' I'm using here, in fact, is not very different from the one Wittgensteinians have in mind when they talk of 'forms of life'. My proposal, then, is to show that observation is intimately linked to background beliefs, habits and actions of the observer. The second step of the overall argument in this chapter will be to explore how this impact of observation on the life of the observer may be expressed in terms of virtues and vices. I will divide the first step into three sections: one on qualities, one on perception and finally one on the extension of the power of our senses. The final section of the chapter will deal with the question of observation and the life of virtue.

Qualities

The main challenge to the view that observation is intimately related to the domain of the practical comes from the idea of radical objectivity, itself dependent on the traditional idea of a clear distinction between the various qualities that we observe. If a clear distinction could be drawn between primary and secondary qualities, it would be possible to define observation of the scientific sort in very simple terms. It would be enough to say that scientific observation is that of primary qualities. These primary qualities are essentially disengaged from any subjective point of view. Observation would thus be understandable without any reference to the life of the observer.

To see whether this argument holds, it is useful to recall the reasons that have been given in the course of history in favour of the distinction. It seems that the origins of the distinction lie in antiquity. Controversy, however, has never been definitely eliminated. Democritus in ancient times had the intuition that things like colour and taste are a matter of convention while what he called the atoms and the void are a matter of truth. In the works of Galileo Galilei and René Descartes, the distinction started to appear in terms of real and unreal qualities. The former were qualities that were considered significant in the sense of deserving the attention of the observer. The latter were qualities that we erroneously think belong to the object while in fact they do not.

The discussion was further refined and developed by a number of later philosophers most notably John Locke. As is well known, the main arguments concern the fact that what happens during the observation of a given phenomenon needs careful evaluation, because the senses do not always deliver aspects that belong directly to the object perceived. John Locke, for instance, in his *Essay concerning Human Understanding*, describes primary qualities as 'utterly inseparable from the body, in what state soever it be' (*Essay* II, viii, 9). He includes as primary qualities such features as extension, figure, and motion. The secondary qualities, on the contrary, are intimately linked to the observer. These include colour, smell and taste. Saying that these are intimately linked to the observer requires further clarification. It brings in the question, for instance, of whether secondary qualities are indeed qualities of the observer rather than of the object. The standard way to proceed at this point is to describe secondary qualities as qualities belonging to the object not directly but in a roundabout way. More precisely, we explain secondary qualities in terms of dispositions of the object to produce sensory experiences in human observers. Locke remains rather vague about the exact nature of this distinction between qualities. As regards secondary qualities, he seems to want to reduce their importance. For him, they are 'nothing in the objects themselves but powers to produce various sensations in us' (*Essay* II, viii, 10). He remains somewhat unclear as to what a power is.

Even though there may be no clear account of a power in this context, there seem to be some reasons to support the claim that a real distinction can be drawn. The argument most readily accessible is probably the one concerning perceptual relativity. The basic idea here is that observers' perceptions of colours, tastes and such like features, as opposed to perceptions of shape or size, are very variable from observer to observer. Locke mentions the example of plunging one's hands in the same container of water. It is a well-known fact that the same body of water sometimes feels hot to one hand and cold to the other, depending on the environment the hands were in before being plunged in. From this experience we learn that hot and cold are not really in the water. The same body of water cannot be both hot and cold at the same time. These two qualities have to be considered therefore sensations that the water produces in the mind. The shape and size of objects, on the contrary, are different from warmth, because they belong to the objects that appear to have them.

If this is right, it would be possible to define observation of the scientific sort by referring only to primary qualities. Moreover, since these primary qualities are essentially disengaged from any subjective element, observation would be understandable without any reference to the life of the observer. This standard way of introducing the distinction between the qualities, however, leaves a lot to be desired. To show that it is not so easy to do away with the observer's life, I will develop my argument in four points.

The first deals with the strength of the argument from perceptual relativity. Does perceptual relativity really distinguish between primary and secondary qualities in the way suggested above? The basic idea was that, from the 'point of view' of the left hand, say, a specific amount of water is perceived as hot. From the 'point of view' of the right hand, the same body of water is perceived as cold. It is however also a well-known fact that a circular table will appear elliptical from the side and circular from above. Hence the Lockean primary quality of shape has an element of perceptual relativity associated with it, making it resemble, to a certain extent, the secondary qualities. The same thing can be said about the other primary qualities mentioned by Locke. If a quality is perceived, it must be perceived from some one particular point. The distinction between primary and secondary qualities therefore cannot be drawn with the required degree of clarity via the argument from perceptual relativity.

Admittedly, a quality like colour is more observer-dependent than shape. The degrees of observer-dependence, however, does not, strictly speaking, show that what are traditionally known as secondary qualities do not belong to the object. Maybe the object has indeed special kinds of properties that allow this kind of great variability. Moreover, the very idea of a primary quality is not without its own inherent problems. A.D. Smith, for instance, has recently uncovered some serious obstacles.[1] He claims that a careful study of the historical development of the debate shows that 'primary quality' is not a straightforwardly philosophical notion, because we have no *a priori* insight into the fundamental nature of the physical, nor, therefore, any such access to the concepts needed for science. He writes:

> my inclination is either simply to ditch the term 'primary quality', or use the term to advert to the properties deemed fundamental by current science. Matters are much clearer

1 A.D. Smith, 'Of Primary and Secondary Qualities', *Philosophical Review*, 99 (1990): 221–55.

> with secondary qualities: those that correspond to our sensorily determined concepts, and which remain, definitely (for us) smells, tastes, sounds, colours and temperatures. (p. 253)

The issue turns on what we take 'fundamental' or 'real' nature to mean. Is it good science because it describes the real? Or is it real because it is what the scientific theory talks about? Smith's proposal seems to be close to the idea that the ultimate notion of 'real' must be acknowledged as dependent on our everyday practice, involved as it is in our constant dealing with secondary qualities. The main point I would like to highlight here is not the one concerning the notion of the real or primary qualities as such. It is rather the issue of the alleged clear distinction between primary and secondary qualities. Smith's study adds weight to the doubts expressed above, because if primary qualities end up as ultimately unreachable, the only currency left would be secondary qualities, at best with various degrees of 'secondariness'. This constitutes my first point in view of showing that one cannot do away with the observer's life. Perceptual relativity has little argumentative weight on its own. It does not do the job of distinguishing between the qualities as well as we thought. The second point concerns perception.

Perception

Observation, scientific or otherwise, certainly involves perception. There are cogent reasons to show that even perception, which, in a sense, is the broader category, entails knowledge, and knowledge, in its turn, is intimately linked to the life of the observer. For perception certainly cannot be accounted for by simplistic explanations that are often mechanistic in character. The seventeenth-century debate on the possible distinction between qualities gave rise to the idea that the world as we perceive it is very different from the world as it is in itself. According to this view, what we apprehend is not the object itself, but rather the ideas that it causes within us. In this way, it was hoped that we could rid ourselves of the problem of naive realism. This position holds that all that perceptually appears to be true of the object is indeed true of that object. It faces obvious problems in accounting for the fact of illusion or hallucination, because it transforms human observers into infallible perceiving-machines, which is not the case. The seventeenth-century debate bypassed these problems by claiming that what we apprehend is something that lies between us and the object, namely the idea of the object. Such a solution, however, cannot do much in the way of persuasion. It cannot stop us sliding unrestrainedly towards the other direction, namely the opposite extreme position that may be called naive phenomenalism. According to this extreme view, *nothing* of what is perceived of an object is true of that object.[2]

2 For a study of various distinctions and locutions related to seeing and believing see R. Audi, *Epistemology* (London: Routledge, 1998), chapter 1; B. Stroud, *The Quest for Reality* (Oxford: Oxford University Press, 2000), chapter 5.

Precisely this kind of sliding seems to be happening in various recent works dealing with neuroscience. It has become customary to consider colours and other secondary qualities as mental constructions created in the brain but non-existent outside it. For instance, Irvin Rock writes:

> The perceptual world we create differs qualitatively from the physicists' descriptions because our experience is mediated by our senses and constructed internally as a representation of the world. [...] What we experience as tones of varying pitch the physicist describes as objects vibrating at different frequencies. Colors, tones, tastes and smells are mental constructions, created out of sensory stimulation. As such they do not exist outside of living minds. The philosopher asks, Does a sound exist when a tree falls in the forest if no creature is near enough to hear it? We can assume that the fall would cause vibrations in the air. They would exist, to be sure. But there would be no sound because a sound, by definition, implies the sensation evoked in a living being by such vibration.[3]

There is a serious misconception happening here that needs to be clarified. The basic starting point for the very use of psychological predicates needs to be rediscovered. Take for instance the verb 'to hear'. When I hear something, who is doing the hearing, me or my brain? As is well known, Aristotle had conceived the soul not as a separate entity from the body but as the collection of all the powers of the living being. This enabled him to avoid making the mistake of attributing to the soul the exercise of, for instance, hearing. The power of hearing is exercised by the living being, not by its soul. In his own words, 'to say that the soul is angry is as if one were to say that the soul weaves or builds. For it is surely better not to say that the soul pities, learns or thinks, but that the man does these with his soul.'[4] The Cartesian picture, on the contrary, ascribes all psychological functions to the mind, and, since the recent trend in neuroscience is towards materialism, many neuroscientists follow Descartes' lead but with the slight twist of ascribing all such functions to the brain. In the views of William R. Uttal, for instance, the materialist trend is presented as indispensable:

> The mind is a function of the brain. [...] This is a fundamental, though often unspoken, axiom of modern psychobiology and to a somewhat lesser degree of the more inclusive neurosciences. Philosophically, this view represents a materialist monism. To assert otherwise is to make a logical mess out of all psychobiological and some neuroscientific research.[5]

The anti Aristotelian use of psychological predicates is evident, for instance, in the reasoning of David Rose: 'The brain is composed of nerve cells, none of which can

3 Irvin Rock, *Perception* (New York: Scientific American Books, 1984), p. 4.

4 Aristotle, *De Anima*, 408^b, 12–15.

5 William R. Uttal, 'Mind, the psychobiology of', in G. Adelman and B.H. Smith (eds), *Encyclopedia of Neuroscience*, vol. II (Amsterdam: Elsevier 1999), p. 1185.

think, see or feel on its own, but when connected together in the right way, they form a brain which does possess such properties.'[6]

This is the source of the misconception. The problem here is logical not empirical. It is logical because the attributes we are dealing with apply only to the whole living being and not to some part of it. Predicates we use to describe human beings come in various kinds. Some can be attributed not only to the whole living being but also to some of its parts. Others cannot. Take, for instance, 'is cold'. This can be said of the whole person, or of some of its parts, such as the feet. Psychological predicates are different because they apply to the human being as a whole. If, instead of saying 'I am seeing the table', we were to say 'my eyes are seeing the table', we would be creating a confusion. We would be implying that my eyes constitute a person in themselves. The mistake here is of shifting the psychological predicates from where they really belong to an imaginary 'small person' presumably situated within the head of the real person.[7] Such a move is completely *ad hoc*, and explanatorily barren, because it creates an infinite regress. M.R. Bennett and P.M.S. Hacker, in their study *The Philosophical Foundations of Neuroscience*, diagnosed this same problem more precisely as a 'mereological fallacy'.[8] The term is derived from the word mereology, which means the logic of parts and wholes. The fallacy occurs when someone applies psychological predicates to parts of human beings, such as the brain or the eyes, when in fact they are intelligibly applicable only to the human being as a whole.

The essential point here is to distinguish between empirical questions from logical ones. For a reply, the first need observation, theory and experiment. The second need careful conceptual analysis. The question concerning the nature of psychological predicates is of the second kind. What empirical study does is explain the process by which we are able to hear, say, the sound of the tree falling in the forest. This it does in terms of pressure waves in the air. This doesn't show, and indeed cannot show, that what appears to us as making a sound doesn't really do so. To examine the question whether there are still sounds in the absence of observers, we need to examine the *meaning* of the psychological predicate related to the secondary quality 'makes a sound', namely the psychological predicate 'hearing a sound'. And examining a meaning is a conceptual task, not an empirical one. It involves studying how the expression is taught, learned and applied to paradigmatic cases. It involves also examining what follows logically from such application.

Here lies the essential link I need for my argument. My hypothesis was that perception is intimately related to the life of the observer. In other words, I proposed that to understand perception, we need to bring in elements that have to do with the

---

6 D. Rose, 'A Portrait of the Brain', in R. Gregory, J. Harris, P. Heard, and D. Rose (eds), *The Artful Eye* (Oxford: Oxford University Press), p. 51.

7 This point is further discussed in A. Kenny, 'The Homunculus Fallacy', in *The Legacy of Wittgenstein* (Oxford: Blackwell, 1989), pp. 125–36.

8 M.R. Bennett and P.M.S. Hacker, *Philosophical Foundations of Neuroscience* (Oxford: Blackwell, 2003), chapter 3.

background beliefs, habits and actions of the individual. This hypothesis can now be seen as justified because of two main reasons presented above. Firstly, psychological predicates are attributable to the living being as a whole. Secondly, their evaluation is a logico-grammatical task not an empirical one. Such a task involves the study of patterns of linguistic behaviour that branch out into the life-world of the perceiver.

This argument may not be sufficient to persuade someone who is a radical revisionist: someone, in other words, who holds that the empirical kind of inquiry carries much more weight than the logico-grammatical one that determines the meaning of expressions, their boundary of sense and non-sense. It is so much stronger that it authorises the scientist to revise the meaning of common expressions. Such an objector would insist that expressions like 'hears a sound' and 'sees the colour red' can indeed be applied to the brain. This new application is a justified *extension*, or even a *substitution*, of the normal everyday use of the terms. And the justification comes from the predictive success of the neuroscientific theories describing the brain.

To such an objector I would reply that some further reflection can readily show how the psychological predicates as applied to a person X are related to X's behaviour in a special way. They are not related to behaviour is a way that needs inductive reasoning for their correct application. They are not related, that is, in a way that obliges the one who is watching X to list the various patterns of X's behaviour, and then *induce* a conclusion regarding X's being in one psychological state rather than in another. On the contrary, psychological predicates are related to behaviour as a matter of fundamental principle. Consider, for instance, seeing and seeing-behaviour. The former is *logically constituted* of the latter. It is not merely manifested by the latter in an empirical way. The link between seeing and seeing-behaviour is thus a logical one not an empirical one. The primary sense of 'seeing' is essentially constituted of the way the living being, as a whole, responds to visible objects, follows them by the eyes, is unable to respond when light is not available, and so on.[9] In other words, what my reply here amounts to is the following point: trying to change the primary meaning of the words that are basic in people's life cannot be done piecemeal because the word and the corresponding behaviour are logically linked to other words and their corresponding behaviour and the entire network will not allow an isolated amendment.

Having achieved some clarification as to the misconception that often occurs in the understanding of perception, let me now go further and explore the epistemological dimension of perception, to explore, that is, how perception is related to knowledge.

9 I am avoiding behaviourism here by avoiding the claim that the meaning of the psychological predicates is *equivalent* to the behavioural description. What I'm holding is that the behaviour is an essential element of the meaning of the psychological predicate. I am leaving space for the possibility of a person pretending to be seeing or to be in pain. The pretending is itself dependent on the primary meaning of the predicates. That is why, to pretend to see something, one needs to open one's eyes, to follow with one's eyes a point in space, and so on; to pretend that one is in pain, one needs to groan, to assuage part of one's body, and so on.

I am here proceeding on the assumption that perceptual knowledge is just one kind of knowledge. There are other kinds, such as knowledge acquired through reasoning or through the testimony of others. These are also ultimately founded on sense perception. It is apparently quite straightforward, therefore, to conclude that perception is also related to the broader horizon of the life of the perceiver via its epistemic dimension. My project of how observation affects moral character would thus appear eminently worth pursuing. Such a brusque procedure, however, would be naive because there is an obvious challenge to overcome. This challenge comes from those who argue in favour of a kind of basic perception that is non-epistemic in character. Their arguments, if valid, would separate perception from knowledge and practice, and thus would undermine the basic assumption behind my project. Fred Dretske, for instance, argues that non-epistemic seeing is a primitive visual ability that is common to a great variety of sentient beings.[10] It is somewhat like what is common to both the eye and the photographic camera when both are situated in front of the same object. This kind of basic-level perception, he holds, is the common platform on which all epistemic ways of seeing are dependent. In other words, if, say, I see something as a table in the epistemic sense of coming to *know* that it is a table, then I need to have had at least a prior *belief* that what I see is, or might be, a table. The level of belief corresponds to the basic-level of perception on which the level of knowledge is dependent. The main reason for his claim is that negating the existence of a level of non-epistemic perception that is independent of all past experience, conceptual schemes, modes of classification and so on would give rise to a kind of radical chaos that is not observed in human society. It would give rise to a situation in which

> people who possessed radically different beliefs, or diverged significantly in conceptual orientation, did not, indeed could not, see the same things. The expert and the novice, the sophisticate and the savage, looking at the same thing, would see different things; and this is but a prelude to the view that we each have a our private perceptual world.[11]

If my reading of Dretske is correct, his main point is that, since people do see the same thing, there must be a non-epistemic kind of perception, detached, as it were, from the conceptual scheme and the modes of classification that partly constitute the life-experience of the perceiver.

On the surface, this argument may seem to threaten my project of showing that observation is intimately related to the life of the observer. There is, however, no such threat here. Dretske is manifestly optimistic about the possibility of having a clear distinction between epistemic and non-epistemic perception. I cannot see how such optimism could be justified. In the normal run of everyday events, no one distinguishes between seeing a thing immediately as an X in the sense of forming a belief, and seeing a thing as an X in the sense of knowing that it is an X. In special circumstances, such as scientific inquiry with sophisticated instruments, such

---

10 F. Dretske, *Seeing and Knowing* (London: Routledge & Kegan Paul, 1969).

11 Ibid., p. 76–7.

distinctions are certainly necessary. It seems, however, largely artificial to project backwards, as it were, from the special case to the normal everyday case, and try to draw a line between doxastic seeing and epistemic seeing in all circumstances.

Moreover, I wouldn't worry too much, as Dretske does, that accepting perceptual relativity even at all levels of perception would lead to private perceptual worlds, each of us seeing what our individual set of background beliefs allows. I wouldn't worry because a private perceptual world is impossible for other reasons, namely linguistic reasons. We see, most of the time, the same objects, in the sense of seeing these objects in common, because the concepts we have are shared concepts. The mental activity of different people proceeds with similar concepts because these individuals need words whose meaning is determined by their common use within the meaning-generating community. These words are used in specific rule-governed ways. It is therefore the public nature of language that guarantees the element of objectivity that Dretske is concerned about. He does not need to conjure up a level of non-epistemic perception.

The considerations relevant to this point about seeing and language are well rehearsed, as regards the more specific domain of scientific observation, in the seminal study of Norwood Russell Hanson in his *Patterns of Discovery*.[12] He shows how there are two very significant influences on observation. These influences effectively change the simplistic view of observation in terms of a passive recording of information to the more realistic view of observation as a form of knowledge. The first influence derives from the fact that there is a theory already present within the mind of the observer prior to the act of observing. Hanson gives the example of the trained physicist and a little child, both of whom behold an X-ray tube. The physicist sees an X-ray tube, while the child a complicated light bulb. He writes: 'There is a sense, then, in which seeing is a "theory-laden" understanding. Observation of *x* is shaped by prior knowledge of *x*' (p. 19). The second influence derives from the inevitability of language. In a radical sense, we need language, or at least some form of notation, to express what we know. Without language, there would be little we could recognise as knowledge. If a person reports that she saw an *x*, we can be sure that she also has some ideas associated with *x*. The *x* must be situated within the logical space of that person. If we see something, we are obliged to see it as one thing of a certain kind. Hanson concludes that 'observation in physics is not an encounter with unfamiliar and unconnected flashes, sounds and bumps, but rather a calculated meeting with these as flashes, sounds and bumps of a particular kind' (p. 24).

In my overall project of showing how observation is intimately related to the life-world of the observer, Hanson's arguments are of considerable importance. It should be emphasised that the project is not to show that there are various kinds of observation, and that the kind we require for science is the one that is intimately related to the observer's life. Rather what I am arguing for, in line with Hanson, is that observation is *always* an act that depends on the broader context of the observer's

12 N.R. Hanson, *Patterns of Discovery: an Inquiry into the Conceptual Foundations of Science* (Cambridge: Cambridge University Press, 1958).

life. There is no clear distinction between the expert and the normal person. Each of these observes according to what is previously present in their background-beliefs. The scientist is therefore not the one who sees and reports in a neutral, machine-like manner, even if that were possible. On the contrary, he or she is the one who sees in familiar objects what the normal person hasn't seen before.

Up to now, I have made two points in view of showing that one cannot do away with the observer's life. The first dealt with perceptual relativity and its alleged role in distinguishing between primary and secondary qualities. The second concerned perception. I move on now to the third point. This concerns the extension of power of the senses. The ultimate aim in this chapter, one should recall, is to see how observation, as a characteristic of scientific practice, affects the moral character of the individual.

Extending the Power of the Senses

Instruments like telescopes and microscopes enable us to perceive what we cannot normally perceive. Such instrumentally enhanced observation has proved essential for the development of science and technology. It is, however, not as unproblematic as it looks. Some reflection on what happens when we extend the perceptive power of the senses is in order. I will proceed by first presenting the main problem in some detail. The main hypothesis I will be testing is an extension of what I was examining above, namely that observation, even enhanced instrumentally, needs to be accounted for with reference to the observer's life.

Let me start with two illustrative historical examples involving instrumentally enhanced observation, one dealing with the extension of the power of the senses towards the very large, and the other towards the very small. My aim here is to highlight the alarming distance that can emerge between the everyday vision of the world and the other image of the world that often arises via the use of sophisticated instruments.

First, consider the question of the motion of the Earth. Our everyday perception makes us believe that the Earth is in a state of rest. We know what being in motion feels like. In a moving vehicle, we experience various forces, jolts, and sometimes bumps. Since none of this is perceived as regards the Earth as a whole, it is understandable how our everyday observation leads us to believe that the Earth is not in motion. In contrast to this, the use of a telescope enables us to have striking proof that we are in fact in motion around the Sun, as famously proposed by Nicholas Copernicus and defended by Galileo Galilei. By the use of powerful telescopes, we can observe the apparent displacement of nearby stars with respect to the background of distant stars. This stellar annual parallax, corresponds to the same kind of observation one can make while watching the countryside from the window of a moving train. Even if the train's motion is imperceptibly smooth, passengers can still conclude that they are in motion. They can deduce this from the fact that nearby trees appear to move across with respect to distant trees. The annual parallax of a nearby star was first observed by F.W. Bessel in 1838, about two hundred years after the Galileo trial. The nearby star 61 Cygni was observed to show a displacement with respect to the distant stars within a period of six months, the time needed for the Earth to go around the Sun from one end of its orbit to the other. As the Earth returns back to its original position, the star's displacement is of an equal amount and in the opposite direction. This is a clear proof, obtained by our sense of vision but enhanced by the use of instruments. It shows that the Earth is in motion. So what are we to make of the everyday perception that the Earth is in a state of rest? Two images of the world emerge. Moreover, since they contradict each other, one needs apparently to be abandoned for the sake of consistency.

The same kind of contradictory conclusions can be arrived at when considering the very small. Atomic theories of matter have a long history, dating from the arguments put forward by Greek philosophers that matter is not a continuum but consists of eternal, invariant, impenetrably-hard, homogenous bits, that move in the void. Developments during the late seventeenth and early eighteenth centuries showed how atomism could be useful to fuse the idea of elements together with the view of compounds. Ironically, the scientific community was convinced of the existence of chemical atoms at the very time when instruments that increase the power of the senses, such as the set-up involved in X-ray crystallography and in studies on cathode-rays, produced ineluctable evidence that atoms are themselves made of smaller particles. In 1911, Earnest Rutherford observed that the great majority of α-particles bombarding gold-plate passed through without any deflection, while some showed large-angle scattering. This is evidence that both the positive charge and the mass of the atom are concentrated in a nucleus which is very tiny in comparison with the atom. We have to accept that, in spite of the ancient Greek views, the main bulk of the atom is empty space. What are we to make of the everyday perception that a normal table is solid? This question has been spelt out most forcefully by Arthur Eddington.[13] His major point is that the picture we obtain via the extension of the power of our senses demands that the table be viewed as constituted of microscopic point-charges in perpetual motion. It demands, moreover, that the total volume of these particles put together be considered less than one billionth of the volume of the table. So, on the one hand, we have the everyday vision of the world involving solidity, while, on the other hand, we have the scientific vision involving empty space. How are we to react to such a conflict between two views of the same thing?

These two historical examples reveal the need to examine the very nature of the two conflicting images of the world. It is perhaps most convenient to use the vocabulary introduced by Wilfred Sellars.[14] He calls the view of the world based on everyday observation the *manifest image*. This is the original image of the world as determined by our being here and now. It includes also a certain amount of refinement

13 A. Eddington, *The Nature of the Physical World* (Cambridge: Cambridge University Press, 1928).

14 W. Sellars, *Science, Perception, and Reality* (London: Routledge & Kegan Paul, 1968).

of ideas that a normal person engages in without elaborate philosophical reflection or sophisticated scientific instruments. Sellars makes the interesting observation that the primary objects of this manifest image are not things but persons. As opposed to this manifest image, we have what he calls the *scientific image*. This is based on the methods of natural science that include correlation between measurable quantities. They include also, as a matter of explanation, the postulation of various entities and relations between these entities. It is important to highlight the point that such correlations and postulations of the scientific image are alien to everyday reflection.

The conflict between the manifest image and the scientific image can sometimes appear as a head-on collision that obliges us to choose between the two competing visions of the world. We are apparently obliged to eliminate one to restore consistency. A number of philosophers of science follow Arthur Eddington on this point and claim that the scientific image should prevail. The issue here is not very different from the one raised above concerning the use of psychological predicates. It was said above that neuroscientists are often tempted, especially if they have eliminativist tendencies, to consider their novel use of such predicates as the correct use, whatever may be their normal everyday use.[15] This is exactly what is happening now on a bigger scale. Is there a flaw in the argument?

The proper reply here is to deny that there should be any conflict in the first place. *A fortiori*, there shouldn't arise any need to eliminate one image rather than the other. Quite simply, the manifest image cannot be eliminated. It plays a central role not only as regards the view of the world we take for granted, but also and more importantly as regards the very meaning of the words we use. In the example of the table, the very meaning of the word 'solid' is generated within our everyday practice. This practice includes perceiving things, engaging in simple movements, making simple predictions of immediate aspects of our everyday living, and so on. The meaning of 'solid' is primarily determined by such everyday common practice. It is a mistake to hold that this meaning derives primarily from some dictionary definition expressed in terms of other notions like 'impenetrable' or 'not made of empty space'. If asked what we mean by 'solid', we can certainly link the notion to other words, as is done in dictionaries. The ultimate response, however, will eventually involve our rudimentary action involved in saying and showing: 'Solid means like this!' There remains therefore an element of analyticity involved in the view I am defending. We are not dealing, however, with an analytic statement that links one concept to another, as we have in 'A bachelor is an unmarried man'. We are dealing rather with an analytic statement that links, if one may say so, a concept to an aspect of rudimentary common, everyday practice. This simple example involving 'solid' can be generalised to cover those words of our vocabulary that deal with

15 Proponents of eliminativist reductionism, such as Paul and Patricia Churchland, conceive of the common everyday explanation of human conduct as 'folk-psychology'. They think that this is a theory that needs to be eliminated so as to give way to the scientific image supplied by recent neuroscience. This project is flawed. I discuss this at some length in: L. Caruana, 'Is Science eliminating ordinary talk?' *Forum Philosophicum*, 4 (1999): 25–39.

the everyday, common experience of humankind. In general we can say that our perceptive faculty, linked as it is to our typically human discursive rationality and with everyday rudimentary practice, lies at the foundation of the meaning-generation of basic words.[16]

To be more precise, I am denying that there can be conflict between the manifest and the scientific images because there are various inevitable relations of dependence between the two images. I need to concede, first of all one important point. Even though the telescope enhances what I do with my eyes, it doesn't follow that the telescope and the eye are similar instruments as far as I am concerned. Some basic principles of perception need to recalled. The one who does the seeing is me as a person. The eye is part of me. This is not the case with the telescope. To construct a telescope and use it properly as an enhanced way of seeing, I need to know some theories about optics, and apply them. This point is more evident when we're dealing with, say, radio telescopes or electron microscopes. It reveals that an observation through an instrument is completely different from simple observation. The former needs a process of inferential reasoning to guarantee that the instrument is working correctly and sometimes also to interpret the results. The latter, in normal everyday circumstances, does not involve any inferential process.

Having conceded this point about a fundamental difference between the eye and the telescope, I proceed now by recalling that simple everyday perception plays an ineradicable role in all observation. To use any high technology instrument for observation purposes, we need to return, somehow or other, to our everyday perceptual capacities. Cloud chambers, for instance, constitute important links between arguably unobservable entities and simple lines on paper humans can see and measure. Geiger counters constitute links between invisible entities and sounds humans can hear and sometimes even count. If the clicks are too rapid, we use another instrument that translates the counter results into a specific swing of a needle, again a swing humans can see. All scientific instruments must ultimately lead to an observation of the kind we have in everyday life: not the other way round. It is absurd to think that we can calibrate our seeing with our theories about microscopic structure of solids instead of the other way round. This is just as ridiculous as the case of me wanting to arrive at a destination without starting here. Just as one cannot go anywhere without starting from where one is, so also one cannot observe without engaging in everyday perception.

It becomes clear now why the onus falls on the scientists to explain why a tabletop is still seen as solid in spite of our belief that it is made up of billions of atoms, the bulk of which is mostly empty space. Scientists do not just propose a new world-view. They must also explain it, describing how it is linked to what we perceive in our daily lives. They do not just propose the existence of atoms that are nearly entirely empty space. They must also postulate the existence of forces between these

16 The central role of common practice for natural science is further analysed in L. Caruana, *Holism and the Understanding of Science* (Aldershot: Ashgate, 2000), chapters 5 and 6.

atoms, forces large enough to produce the macroscopic effects we see when dealing with the table in our daily lives. Similarly, the onus falls on the scientists to explain why we on Earth have the experience of being in a state of rest in spite of the fact that the Earth rotates on its axis and also revolves round the Sun. They have to add an explanation as to why we do not have the experience of motion as we do when, say, on a moving boat, or during an earthquake. One recalls here that it was precisely the obligation to search for this explanation that gave birth to Galilean Relativity theory.

Discussing the manifest and the scientific images of the world, Sellars writes: 'the conceptual framework of persons [i.e. the manifest image] is not something that needs to be *reconciled with* the scientific image, but rather something to be *joined* to it.' (p. 40) This is an essential task of the entire enterprise we call science. The world depicted by the manifest image, and the one depicted by the scientific image, are both the same world, our world. Neither one nor the other should be considered a problematic alien appendage. People like Eddington and the eliminativists in cognitive science are grossly mistaken in discrediting the importance of the manifest image.

I need not rehearse here the arguments that have done increasing justice to the manifest image during the twentieth century especially under the influence of the later works of Ludwig Wittgenstein. Sellars adds considerable weight to the point I'm urging by recalling that the analytic tradition in Anglo-American philosophy 'has increasingly succeeded in isolating [the manifest image] in something like its pure form, and has made clear the folly of attempting to replace it *piecemeal* by fragments of the scientific image' (p. 15). There can be no question of elimination or of replacement of some basic terms of everyday practice so as to be more in line with recent scientific findings. It is not difficult to see how a clear category mistake occurs every time someone mixes up discourse dealing with things we see, like Eddington's table, with discourse dealing with things we don't see, like molecules. It is correct to say 'The table is brown', if it is. It is absurd, however, to say that the molecules are brown. We may perhaps say the system of molecules *looks* brown. But to say that the system of molecules *is* brown would generate confusion. It amounts to a category mistake, because the expression 'table' can readily enter the sentence-form '... is brown' but the expression 'system of molecules' cannot. If inserted into that sentence-form it would create the same kind of absurdity as the one arising from, say, the sentence 'Aristotle is a multiple of 3'. The absurdity here occurs not because of a lexical or grammatical irregularity but because of the misguided effort to combine the logically uncombinable. Predicates have a logical space determined by their meaning. This space needs to be respected. And to know the extent of this logical space, one can revert only to the way the predicate is used in practice within the complex network of uses of other predicates and words.[17] The upshot is that

17 I am drawing inspiration here from Fred Dretske's treatment of this point. He explains a category mistake by analogy with what happens during a play. He writes: 'to cross a "category" boundary by saying that the system of molecules *is* red is to generate the same sort

the manifest image is not only ineradicable, but also of central importance as the ultimate point of reference for instrumentally enhanced observation.

This is a convenient place to summarise the argument I have been presenting up to now. The ultimate aim of this chapter is to explore how observation, as part of scientific practice, may affect the moral character of the individual. My first step was to defend the idea that a genuine account of observation cannot disregard the essential link of this activity to the life of the observer. It is useful to recall that by 'life' here I mean the combination of background beliefs, habits and also the particular actions of the observer that, in the long run, may enhance or impede the development of some of these habits. At face value, observation is easily classified as a feature of the individual that has to do with theoretical wisdom. Habits and action, on the contrary, are often considered features that have to do with practical wisdom. The challenge in my proposal, therefore, involves unearthing the link between these two dimensions of the same one individual. I have presented my case in three steps. The first point dealt with perceptual relativity and its alleged role in distinguishing between primary and secondary qualities. The conclusion here was that the distinction between these qualities is not clear. The second point concerned perception. I showed that it would be a mistake to claim that we need to assume the existence of non-epistemic perception. Observation is, as it were, thoroughly theory-laden. The third point was about the particular questions that arise because of instrumental enhancement of the power the senses. In this last section, I argued that the manifest image of the world cannot be eradicated by the scientific image derived from sophisticated observation. The central, ineradicable role played by the manifest image shows that observation, even if enhanced by instruments, is a feature of natural science that needs to be understood with reference to everyday practice.

Observation and the Life of Virtue

So far, I have been concerned with the nature of observation. It is now time to address the question of the way such scientific observation could affect the habits of the individual. There is need here to spell out the steps of my reasoning. The basic result of the previous sections, namely that observation cannot be adequately accounted for if not in reference to the observer's life, can be expressed in a fuller form by saying that a full account of observation needs a reference to the linguistic habits and form of life of the observer. I am not pretending that this constitutes a completely original thesis. If there is any question about originality, my contribution here should be seen as limited to the extension, as it were, of the well-known thesis defended by Hanson, namely that observation is theory-laden. In the version I am defending, this thesis is extended to cover not only an influence from background theory but also from everyday practice. One may emphasise this point by saying that observation is not only theory-laden but also action-laden.

---

of confusion as is generated by crossing of the boundary between people and the roles they play [in a Shakespearean drama]' (*Seeing and Knowing*, p. 255).

The crucial step in my argument occurs here. Hanson's thesis may be described in terms of a reciprocal relationship. His point is that observation is as sensitive to background theory as are theoretical developments to the observations on which they depend. My extended version may be described in the same reciprocal way. Observation is as sensitive to theory and action, in other words to the life of the observer in the sense I've been employing so far, as are theory and action on observation. This reciprocal relationship reveals the interesting fact that, in some specific ways, life is observation-laden.

Now my task is clear. I am setting myself the assignment of revealing through analysis some significant features of the way scientific observation as a skill, in so far as it can be seen as the ability to perceive more in things than the normal person, affects certain habits of the individual. My special concern will be the way it affects the ability of the individual to stand in a proper relation to what is good. By this last expression, I mean the ability of the individual to act in line with genuine human flourishing. The moral dimension comes in here. Before proceeding beyond this crucial step, I need to dwell on two points in view of setting the boundaries of the inquiry.

First of all, there is certainly an enormous variety of habits related to observation that could be the object of such an investigation. There is therefore the danger of spreading oneself too thinly over a vast area with very little significant results. It seems reasonable to concentrate on one particular kind of habit and one particular aspect of observation. As mentioned above, the habits that will be considered here are not those dealing with particular tasks of immediate import on the life of the individual, more properly called skills, such as being good at remembering. They are rather those habits that deal with doing good, in a deeper sense. The aspect of observation that presents itself as particularly significant for the study of these habits is the area dealing with science and ordinary language, in other words, the area dealing with the interaction between the scientific and the manifest images of the world.

The second point to be emphasised concerns moral discussion. It is crucial to recall that anyone engaged in arguments on moral issues, that is on issues dealing with matters like goodness, virtue, righteousness, praisworthiness and so forth, is bound to assume certain fundamental moral facts as a common ethical platform on which the discourse can be constructed and sustained. As I mentioned in the first chapter, I am assuming the existence of some common ground as regards the evaluation of certain basic character traits of persons. I am assuming, for instance, that everyone is in favour of such virtues as charity, honesty, justice, and courage. I am assuming that everyone is against hatred, dishonesty, cruelty and cowardice. How these terms are to be understood in special circumstances may need refined argument. Initial agreement, however, as to their quality of being beneficial or detrimental to human flourishing is essential.

Enough has been established then to enable a reasonably situated inquiry into the way the life of virtue is observation-dependent. Consider a typical person living before the scientific revolution. Such an individual would be untouched by many scientific ideas that we now take for granted. We may reasonably describe such a person as possessing only the manifest image of the world. The manifest image supplies the entire action-space for that person. Such a person is a fiction, certainly, and the situation quite a counterfactual one. There was no definite moment when the scientific revolution can be said to have started. Even before the impact of the revolutionary work of Nicolas Copernicus and Galileo Galilei, some elements of the scientific image were present all along the course of history. As an ideal type, this typical person, untouched by the scientific image, is nonetheless useful for our inquiry, especially because contrasting between such a person and another one who is indeed living after the scientific revolution is illustrative. This latter individual does not only inhabit the manifest image but is faced with two world-views, the manifest image and the scientific image. While the first ideal person engages in action within one relatively clear and homogenous context, the second engages in action in a two-layered world. The first individual, for instance, engages in a world whose primary units are persons. The second, engages in a world whose primary units fluctuate between persons and organised agglomerations of particles.

It is precisely this fluctuation that reveals the possible effects scientific observations, as a skill, can affect the attributes of the person as regards doing good. I proceed with the assumption that particular skills, which are styles of action, as it were, affect the individual's mental set-up, have an effect on the dispositions of the person involved, and hence may have a bearing on that person's virtuous life. Following the lead from Aristotle's method, two extreme positions will be identified as leading the person astray. These will correspond to two opposing vices. Once these vices are determined, the nature of the virtue associated with observation will be discernible as the mean between these two extremes.

The first vice is associated with the tendency of resolving the fluctuation between world-views by systematically believing that the scientific image is the correct one. This belief has some benefits. It allows the individual, for instance, to retrieve a sense of unity of perspective. It obliges, however, the discrediting of the manifest image. It is my contention that this habit leads the individual to act in a way that is detrimental to genuine human flourishing. In other words, my claim is that this habit can be classified as a vice. Some reasons are called for here. I need to show *why* this way of acting is a vice. It may seem that such reasons are not really needed. After all, Aristotle himself doesn't seem to have asked the fundamental question 'Why is cowardice a vice?' He takes this for granted. For him, that cowardice is a vice seems to form part of the indispensable platform of moral starting-points. Cowardice is a paradigm vice, as it were. In this, I think, he is right. When one is dealing with other tendencies or habits, however, some argument is in order. The most promising ways of arguing on such an issue seem to be two. The first way is to show that the habit in question goes against human nature as rational, and thus against the very idea of genuine human flourishing. This way of arguing would involve showing how the habit in question causes, for instance, such states as confusion, disunity, fragmentation, progressive self-annihilating, death. The other way is to show that the habit in question is another form of a well-known vice.

The habit in question here is the one of acting systematically on the belief that the manifest image is mistaken, or inconsistent, or unnecessary. This habit can manifest itself in various ways, the most consequential of which perhaps would be when the person will show the strong desire to justify everything scientifically. This situation may be described as a kind of craving for scientism. At the same time, however, the same individual will be obliged to live with the realisation that the manifest image cannot be eliminated. It is always present and ultimately indispensable, as explained above. A persistent sense of internal division within that person will therefore result. The conflict that allegedly exists between the two images infests the integrity of the person. Such a state has negative effects on that individual's action in so far as it reduces his or her capacity to live a unified life. In a nutshell, the conclusion can be put as follows. If scientific observation fascinates the individual to the extent that the manifest image is considered fraudulent, that individual is hindered in his or her project of living in a genuine human way. We are dealing here, therefore, with a vice.

Having fixed one pole of the Aristotelian triad, as it were, we may proceed by investigating how there can occur, within the individual, the opposite tendency. This tendency is associated with the exaggeration of the importance, not, this time, of the scientific image, but of the manifest image. The habit here can be described as the individual's disposition to act systematically in the belief that the scientific image, with all its paraphernalia of theoretical entities, is a pure fiction, while the manifest image is the correct one, in the sense of being the only one that counts. The agent we are considering therefore is one whose action is determined by what is often called instrumentalism. As a philosophical position, this claims that the scientific image is not an image at all. It is certainly not a description of the world. It is rather a system of symbols that enable us to predict certain outcomes that make a difference to what we observe in our everyday life. Ernst Mach's positivistic approach, especially in the way he valued operational definitions for scientific terms, can be considered a good representative of those in favour of instrumentalism. Consider, for instance, the scientific term mass. People who follow Mach will speak not of mass but of the concept of mass. Within the vocabulary of the scientific image, they define the concept of mass operationally in terms of physical and mathematical operations that can be carried out in the laboratory or in the field. For them, therefore, the major terms constituting the scientific image are not considered as having any referent but as significant only in so far as they are identical, by definition, to some aspects of the manifest image. In the case of mass, the concept is identical to the description of the operation involving a particular experimental set-up, such as a beam-balance.

This position has been attacked on various fronts. The major problem with it is probably that it does not allow any distance between the operation and the concept. In the case of mass, when the operation of weighing is equivalent to the concept of mass, there cannot be any question of asking whether a particular weighing operation is functioning properly or not. This is an issue that I will side-step. My main concern is not to see whether the thesis proposed by operationalists and instrumentalists is self-consistent or not. My concern is rather to see whether the *practical* dimension of the individual operationalist or instrumentalist is affected positively or negatively by that thesis. The link between the theoretical and the practical, as regards this point, is reflected in the similar link between scepticism and its practical counterpart, classical Pyrrhonism. At face value, scepticism is often taken to be an epistemological position. Pyrrho however makes of this epistemological position a way of life, essentially centred on the aspiration to attain imperturbability. The price he pays for this state is the systematic disengagement from all inquiry about the deep nature of things. Defenders of instrumentalism, those who work with operational definitions and disdain the scientific image, are in the same position as Pyrrho. They mistrust to the extreme our capacity of using reason and experiment to attain a description of the world that goes beyond the manifest image. If their action corresponds to this mistrust, their tendency would be to act in a way that neglects the full potential of their own intellectual abilities. Such a tendency amounts to self-annihilating tendencies. There is enough evidence therefore to see this tendency as going against genuine human flourishing, and is thus vicious. It corresponds to a kind of false humility.

This vice together with the previous one constitute two opposing dispositions. The first one gives rise to action in line with too much importance given to the scientific image. The second one gives rise to action in line with too little importance given to the scientific image. The first one is a kind of proud overestimation of reason and experiment. The second one is a kind of false humility, an underestimation of reason and experiment.

Virtue lies in the middle. As regards the area covered in this chapter, it seems reasonable to conclude that virtue lies in acting within the perimeter of a right balance between the scientific and the manifest images. On the one hand, even though some of our commonly held beliefs may be mistaken, the virtuous person recognises that they constitute the inevitable point of departure for any discovery, and acts accordingly. The relatively stable set of beliefs concerning our rudimentary everyday life in the world, the set of beliefs we call common sense, is the stable infrastructure needed for any scientific inquiry to start. Novel philosophical ideas and surprising scientific discoveries are always encountered against the background of this common starting point. On the other hand, even though some of our scientifically held beliefs may be mistaken, the virtuous person recognises that they constitute the results of the exercise of our intelligence, and acts accordingly. The scientific image may need constant refinement, just as much as the manifest image does. The virtuous person would be the one who acknowledges the central role played by common sense and common practice, admits the possibility of error within the implications of this set of beliefs, and hopes that natural science or some other discipline will help remove prejudices and gain more true beliefs.[18]

18 This virtue corresponds to the position called critical commonsensism developed by Charles Sanders Peirce. His attention was on this position as a theory of discovery and not, as I'm presenting it here, as a way of describing character traits of an individual. See his 'Pragmaticism and Critical Common-Sensism' in C. Hartshorn, P. Weiss and A. Burks (eds), *Collected Papers*, vol. 5 (Cambridge, MA: Harvard University Press, 1931–1958), pp. 346–50; 'Consequences of Critical Common-Sensism', in Ibid., vol. 5, pp. 351–75.

Chapter 3

Method

To facilitate the identification of virtues and vices that can be affected by the scientific method, it is essential that the main features of this method be first well understood. What scientists actually do is, to a large extent, governed by rules of method. Some of these rules are formulated explicitly, but not everything that is done in science falls under some explicitly formulated rule. Scientific method needs to be distinguished from science itself. As was highlighted in the previous chapter, I'm taking science to be more than a merely theoretical exercise. I'm considering it a complex activity comprising both an intellectual and a practical dimension. A study of method can therefore be seen as a second level reflection with respect to the actual actions carried out by the individual scientist in specific circumstances. Scientists reflect on what needs to be done here and now. Philosophers of science reflect on regular patters of scientists' behaviour that guarantee success. The method is the unifying link, as it were, between the various individual acts. There is little doubt that, when people speak of the scientific mentality, an essential part of what they mean is associated with this method. One rightly expects the scientific method, via the mentality it generates, to have a considerable effect on the attributes of the individual scientist. The most general aims of this method are sketched in section one of this chapter. This is meant to set the scene by presenting a somewhat idealised situation. In section two, the inquiry proceeds with an exposition and evaluation of the logical backbone of scientific discovery, namely induction. My basic aim here will be to bring out the essential links between induction and everyday, rudimentary, human action. The same aim will determine my line of reasoning also in section three dedicated to falsification. In both cases, I hope to build a case for the claim that a full understanding of both induction and falsification cannot neglect this essential link between, on the one hand, the pure logical discourse and, on the other hand, what people typically do in everyday situations. It is from this inevitable link that I will draw some insights concerning the scientific mentality and its effect on the attributes of the person.

Preliminary Description of the Objective

The content of our experience changes continuously. Given this fact, the best kind of knowledge to aspire to seems to be knowledge that goes beyond such fluidity, itself unchanging when all else changes. This kind of knowledge can be called objective in the sense of being knowledge of what is necessary and universal.[1] By knowledge of what is necessary here is meant a knowledge of what does not depend on a human, subjective point of view but is, ideally, common to all points of view. This can be called knowledge of the mind-independent world. The world of experience, although always changing, shows a remarkable dimension of regularity. There are some laws that go beyond human conventions and customs, even though natural laws and human conventions are very often intricately linked. For instance, the fact that water boils at a hundred degrees centigrade is a fact that combines both a necessary element, namely the fact that water boils when heated, and a conventional element, namely the assignment of 100 to that particular point on the Celsius scale. The best we can hope for in our scientific work is to arrive at a knowledge of facts of the first kind, facts like the one that water boils when heated always at a certain temperature, whether our temperature scale is in degrees Celsius or degrees Fahrenheit, or in any other version. The necessity associated with this knowledge is not of a logical kind. It is not derived from the laws of logic, as for example the necessity inherent in the statement: 'It is either raining or not raining.' The necessity with which scientific knowledge is ideally associated is of a nomic kind. This means that its source is found in the laws of nature, in the regularity and order that exists in the world independently of human beings. As living beings, we are indeed considerably dependent on at least some knowledge of this regularity and order.

Scientific knowledge is objective, therefore, in this sense of being knowledge of what is necessary. It is also objective in the sense of being knowledge of what is universal. This term refers to a kind of knowledge that is independent on any temporal or spatial localisation. It is knowledge of what is independent of time and place, and is valid for all entities in a certain category. Talking of such knowledge of what is universal expresses the hope is that prolonged reasoning based on experiment will enable us to purify our common-sense beliefs by distinguishing between aspects that depend on the individual's localised point of view, and aspects that do not. To the extent that this is possible, one arrives at a kind of knowledge that is valid for all rational creatures at all times and places. The crucial assumption being made here is that, by removing the aspects that depend on a particular point of view, one after the other, one will indeed find a precious kernel that satisfies our initial aspiration. Much of what happens in our everyday life is evidence that such an assumption is very reasonable. We consider simple knowledge claims as valid for all people. To increase our stock of knowledge of what is universal, we require experiments to be repeatable by other individuals situated at different places and different times. Any affirmation that lies outside the possibility of ever attaining the status of this kind of universality, lies *ipso facto* outside scientific discourse. Hence, a discourse

---

1 On objectivity of judgement, some basic arguments are presented by I. Kant, *Prolegomena to any Future Metaphysics that will be able to present itself as a Science*, trans. P.G. Lucas (Manchester: Manchester University Press, 1953), § 19. These are discussed in R. Meerbote, 'Kant's Use of the Notions "Objective Reality" and "Objective Validity"', *Kant-Studien*, 63 (1972): 51–58.

that parades itself as being universal and yet does not allow repeatability in its legitimating procedures, as happens for instance in the domain of parapsychology, cannot be considered scientific.

It should be emphasised that the aim in presenting this quick sketch of the objective of natural scientists is not at all to argue that scientific knowledge is indeed constituted entirely of what is necessary and universal. My intention is rather to offer a rough, preliminary idea of what is sought by scientists, with uneven success. It is indeed true that significant amounts of what is said in, say, biology and palaeontology is not repeatable. This fact, however, does not mean that such claims of biology and of palaeontology lie *in principle* outside the realm of repeatability. The universality and necessity with which natural science is associated are best seen as constituting an ideal. An investigation of the ideal case is useful in so far as it determines the characteristics of what the discipline aspires to be. As will be discussed in later chapters, this ideal picture differs from what science actually is in the complex interaction between theory and practice. Nevertheless, the suggestion here is to proceed by starting with an inquiry concerning the ideal case. This will supply the necessary background against which the deviation from the norm could, later on, be better evaluated.

Induction

Drawing a conclusion from given premises can be carried out in different ways. Induction is one of them. In the simplest situation possible, it involves extrapolating from experience so as to arrive at a general statement. Even in this simple form, it has given rise to a number of arguments and counter-arguments put forward by many generations of philosophers. My aim here is not to give an overall survey of these arguments. It is rather to highlight one particular dimension of the issue, namely the necessity of involving the acting person within the discussion.

As the ancients had observed long ago, induction and syllogistic deduction are different but related ways of inferring a conclusion from some given premises. In Plato's *Republic*, book VI, we already find the idea of induction couched within his suggestion that there are 'two kinds of thing, visible and intelligible' (509^d).[2] Plato further elaborates this by saying that the visible can be divided into images and the originals of these images. Likewise, the intelligible can be divided into two directions of cognition. The first direction involves the knower going *to* the forms just like the student of geometry who goes beyond the visible figures to arrive at the pure figure, say the circle itself, or the square itself. The second direction involves the knower proceeding *from* the forms. Reason here starts by grasping the unhypothetical first principle. It then 'reverses itself and, keeping hold of what follows from it, comes down to a conclusion without making use of anything visible at all, but only of forms themselves, moving on from forms to forms, and ending

2 Plato, *The Republic*, trans. G.M.A. Grube, in J.M. Cooper (ed.), *Plato: Complete Works* (Indianapolis: Hackett Publishing Com. Ltd., 1977), pp. 971–1224.

in forms' (511[c]). There can be little doubt that Plato's insights on these movements within the act of knowing are a precursor of the now famous logical distinction between induction and syllogistic deduction. Aristotle retained the distinction between the two directions within the thinking process. This can be appreciated by recalling the distinction between induction (ἐπαγογη) and reasoning (συλλογισμός) in his *Topics*. Of reasoning, he writes: 'reasoning is an argument in which, certain things being laid down, something other than these necessarily comes about through them' (*Topics* 100[a], 25).[3] This corresponds to Plato's 'coming down to a conclusion'. Of induction, Aristotle writes: 'induction is a passage from individuals to universals, e.g. the argument that supposing the skilled pilot is the most effective, and likewise the skilled charioteer, then in general the skilled man is the best at his particular task' (*Topics* 105[a], 15). Here he is effectively noting that the predicate 'is skilled' is concurrent with the predicate 'is effective in his task' for the two mentioned cases. Aristotle then generalises for all cases of men who are skilled, but who are different from each other as regards other aspects. He is therefore engaged in moving towards the universal, as it were, and not away from it.

From these origins of the controversy, we can already notice the fundamental difference between induction and deduction that explains why natural science is intimately linked to the former. Consider the following classic example of a syllogism. We accept that all men are mortal, and that Socrates is a man. We deduce that Socrates is mortal. In such an inference, the conclusion is drawn without any shadow of doubt. If the premises are true, so will the conclusion. This makes syllogistic deduction very attractive. Doubts may arise, however, as to their heuristic value. It is not difficult to see that the general premise, all men are mortal, already contains all the information we obtained after having gone through the syllogism. This shows that a deductive syllogistic argument, even though it may be said to sharpen our appreciation of what our general principles really mean, does not lead to new knowledge about the world. All the conclusions one can ever draw by deduction are already there, hiding within the premise.

Induction is different. From a finite set of particular cases, we arrive at a general statement. There is an acquisition of new knowledge because, on observing, say, only some swans, induction allows us to say something about all swans — even those we haven't observed yet. The problem starts becoming apparent: does induction really allow me to make this jump from some observations to all? In David Hume's famous words, 'we have no reason to draw any inference concerning any object beyond those of which we have had experience'.[4] The aspiration of science to arrive at general claims, valid for all times and for all places, is undoubtedly very ambitious. It seems reasonable to have serious doubts whether induction could ever guarantee success in achieving this aim. Strictly speaking, if we want to make a

3 Aristotle, *Topics*, trans. W.A. Pickard-Cambridge, in R. McKeon (ed.), *The Basic Works of Aristotle* (New York: Random House, 1941).

4 D. Hume, *A Treatise of Human Nature* (London: Fontana, 1970–1972), Book 1, Part 3, sec. 12.

claim that is valid for all times and places, as say 'All swans are white', then we have to check it for all times and places. And this is certainly impossible. Even if we have a very large number of observations to start with, we cannot be completely sure that whatever we want to claim about them holds for all times and places. One may perhaps want to argue that, if we have more modest aspirations, we may be justified in claiming that we obtain new knowledge by induction. Hence, for example, if we have a finite number of possible observations, then we can avoid the frightening concept of infinity that creeps in whenever we talk of 'all times and all places'. Suppose that there exist only a hundred swans. Having seen that ninety nine are white, I use induction to claim that all swans are white. Am I thinking rationally? An affirmative reply here does not seem to be possible. The hundredth swan may indeed be black. Hence, induction seems to be a flawed way of thinking. Hume's point is that anyone who uses induction is not thinking rationally. The scientific mentality seems to be seriously contaminated with irremediable insecurity.

To launch a first objection against this rather alarming claim is not difficult. I can argue in the following way. I have used induction many times in the past, and it proved itself a reliable way of thinking. Admittedly, I make the occasional mistake. But, by and large, induction has proved its worth all along my day-to-day living, and indeed throughout history. This shows that I am entitled to use it again. No one should accuse me of being irrational when using it. This spontaneous defence of induction, however, will not impress hard-line critics.[5] They will insist that this objection is not valid because it uses an inductive method to prove the validity of the inductive method itself. The way the objection goes, in fact, is equivalent to the making of a list of cases of induction in my previous experience: I_1, I_2, I_3, and so on. Each of these was useful and desirable – so far, so good. From these observed cases of induction, however, we are not entitled to draw the conclusion that *all* cases of induction are useful and desirable, because drawing that conclusion would be using the very principle whose validity we want to prove. What opens up here, obviously, is the question of various levels of induction, recently investigated by Brian Skyrms.[6] For the first level, one quantifies over individuals, for instance swan number one, swan number two, and so on. For the second level of inductive argument, one quantifies over the arguments of level one. As an example, consider an inductive argument concerning swans, and an inductive argument concerning ducks, an inductive argument concerning chickens, and so on. This level corresponds to what Hume is talking about when he realised that we are tempted to justify induction by resorting to induction. Skyrms recognised that there are rules for assigning inductive probability for each level. This fact, however, doesn't in itself count as a vindication of induction. The effectiveness of Hume's attack is present in a new guise, because in Skyrms' analysis there is no guarantee that higher level induction is less problematic than lower level induction.

---

5 D. Hume, *An Enquiry concerning Human Understanding: A Critical Edition*, ed. Tom L. Beauchamp (Oxford: Clarendon, 2000), IV, i, 32.

6 Brian Skyrms, *Choice and Chance: an Introduction to Inductive Logic* (Belmont USA: Wadsworth, 4th ed., 2000).

There is no level at which we encounter the same kind of necessity that characterises the deductive syllogism. People like Hume will remain unsatisfied.

So the prospects of building science on the firm foundations of induction look very bleak. Even simple generalisations, like 'all swans are white', do not merit our firm acceptance, because, as we know, a black swan from Australia may be brought in as the one counterexample that shows that our jump from the observed cases to the general rule was not justified. Nelson Goodman's discovery of new versions of the problem of induction complicates things and thereby seems to place natural science on ever weaker logical foundations.[7] His point can be appreciated by considering the following additional questions concerning the swan example. On what grounds do we say that the black bird brought from Australia is indeed a swan? Doesn't the very fact that it is black show that it isn't a swan? These questions indicate that the problem of extrapolation runs deeper and is more complicated than Hume suggests. If we concede that our future observations of swans will be similar to our past observations of swans, we still have to decide which aspects of the birds will indeed show this similarity. It seems arbitrary that we usually take the similarity to be in the colour of the swan rather than, say, in its size.

What kind of reaction can one have when faced with these arguments that apparently undermine one of the most fundamental aspects of our thinking? Hume's famous suggestion was that the way to explain why we make the inductive jump is to say that we have a certain habit. Every time we see the constant conjunction of two events, together with their spatial contiguity and the same temporal succession of one after the other, we have the habit of claiming that the two events are necessarily connected. Since habits are not an essential attribute of a person, in the sense that they can be gained and can be lost, Hume's proposal implies that it is possible in principle to meet people who are like us in all things except in not having this habit. We, who have this habit, and haven't shaken it off, may recognise that it is often satisfying and even useful. What we cannot say is that inductive logic forms an integral part of rationality. In line with Hume's attitude, to know what rationality really involves, we need to look at the syllogism. The syllogism sets the standard.

Hume's treatment of induction has been the source of vast amounts of literature. What I propose in the following paragraphs is to limit the choice of arguments in line with my original aim. I intend to uncover the significant role that the acting person plays in the very evaluation of inductive logic. I will proceed therefore with some arguments dealing with two aspects: firstly, some arguments that show how Hume's approach is fatally reductionistic, and secondly, some other arguments that are related to the so-called ordinary-language defence of induction. The purpose here is to highlight the important links between induction and everyday human action. These links will be the source of my eventual exploration of how the scientific mentality, as represented by the methods of science, may affect the moral character of the individual.

---

7 See Nelson Goodman, *Facts, Fiction and Forecast* (London: Athlone Press, 1954), chapters 3 and 4.

As an initial clarification, it helps to see Hume's reductionism in two ways. I will call the first one benign reductionism, and the second malignant. The benign sort is present whenever induction is analysed in terms of a list of propositions from which a general statement is derived, with no reference at all to the complexity that is involved in real life situations. When the logical problem is transferred to the world of experience, the adjustments that need to be made are so important that the very formulation of the logical problem comes out as very artificial. It is precisely this artificiality that renders Hume's approach reductionistic. He abstracts too much from real life situations, and ends up with a misleading caricature. Consider for instance the normal presentation of the logical structure of enumerative induction as a series of singular claims from which a general claim is drawn. Such a presentation neglects the semantic context needed to make the singular claims in the first place. This neglect involves a kind of abstraction that has famously been placed on a pedestal as a guiding ideal by Francis Bacon, when he suggested that all preconceived ideas about the phenomenon under investigation need to be discarded so as to arrive at a pure inductive procedure.[8]

The problematic element of this kind of reduction here becomes evident when we recall that real-life situations are much more intricate than the text-book case of enumerating statements about the colour of swans. Suppose, for instance, we want to understand malarial disease, and we therefore engage in an inductive procedure to grasp its causes. We usually start with a number of features that seem relevant to the causal chain leading to malaria. Hence we may think of focusing our attention on three aspects: (A) the presence of mosquitoes; (B) malnutrition; (C) humidity. We then apply John Stuart's Mill method of difference and proceed with selective observation in the following sense. If we have all three factors together, ABC, we observe the disease. If we remove one factor, and have only A and B, then we also observe the disease. If we have only B and C, however, we do not. Hence, factor A must be relevant within the causal chain leading to malaria.[9] This procedure is quite standard. It is, however, entirely dependent on our prior selection of the factors A, B and C. In the great majority of cases, to arrive at a set of possible relevant factors for a particular effect is an operation which is totally dependent on our prior assumptions. We cannot start our observations without having some reasons, or at least some prior mental tendencies, that make us choose one set of factors rather than another. Assuming that inductive procedures can be analysed without any reference to such prior sets of reasons or tendencies is very artificial.

Those who still insist that observation should be carried out with no prior assumption whatsoever may want to object here. They may want to argue that, when engaging in any scientific inquiry, we should simply record everything. Only

---

8 Francis Bacon, *Novum Organum* trans. P. Urbach and J. Gibson (La Salle, IL: Open Court, 1994), sections 38–68.

9 John Stuart Mill in his *A System of Logic, Ratiocinative and Inductive: being a connected View of the Principles of Evidence and the Methods of Scientific Investigation*, ed. J. M. Robson (Toronto: University of Toronto Press, 1978), Book III, chapter 8, § 2.

then will we be true scientists, as opposed to opportunists trying their luck and guessing their way through life. This objection however is simply unacceptable. The simple truth is that one cannot record everything. According to such objectors, when seeking the causes of malaria, before doing any experiments, one should continue the list A, B, C, D, E, ... indefinitely. But this will block all kinds of research because the list will never end: it will come to include literally the entire universe. A good dose of imagination that categorises the universe into relatively autonomous packets is an essential ingredient of all scientific work. Only with such a proviso can we understand how a scientist like Isaac Newton worked. Only if we admit the importance of scientific imagination can we appreciate how he managed to put together observations like the falling of an apple, the effect of the moon on the tides and the motion of the planets round the Sun, three observations that are seemingly completely unrelated. He selected these phenomena, and only these, so as to show that they are, in fact, instances of the same general principle.

The upshot is that Hume's approach tends to see induction as a much reduced version of what is involved in the drawing out of a general conclusion from experience. Considering the problems of induction with reference only to the enumerative structure of propositions concerning the same kind and the same attribute neglects some factors that are essential in understanding the logic involved in natural science. Having come to this conclusion, however, I will concede that this kind of reductionism is best considered benign, in the sense that followers of Hume can still retain their ground. They can still insist, correctly, that if the reduced version of induction they extract from scientific practice contains inherent problems, there is something to worry about. The fact that they slough off important factors to arrive at the version they analyse does not directly invalidate what they say, especially if they limit their comments to the logical structure they deal with.

The other kind of reductionism, the kind I'm calling malignant, is more serious. There is a very important assumption being made by followers of Hume whenever they present their conclusion that inductive logic falls short of the acceptable standard of rationality. They assume that it is the syllogism that sets the standard. They are, as it were, infatuated with the deductive necessity exemplified by the syllogism. If the logical problem of induction is formulated in a way that exposes this assumption, it is exposed as a problem created by our own initial restrictions on the term 'rational'. We started with the implicit premise that, in drawing conclusions, rational people need to limit themselves to those conclusions that are necessarily true. This implicit premise essentially means that only syllogistic deduction qualifies as a rational mental operation. Then we noticed that induction is certainly not a syllogism, and from here we conclude that induction does not qualify as a rational mental operation. This argument however is completely dependent on the implicit premise – which is hardly plausible. The implausibility becomes obvious when we recall that 'rationality' is a word with a certain meaning, and the meaning of words is not stipulated once and for all by some assertive philosopher. It depends to a large extent on their use in a living language. One adopts the premise that only the syllogism is rational, perhaps to give the impression of being precise or rigorous.

This inevitably leads to an unacceptable result, as one should expect. It leads to the surprising conclusion that all our everyday practice is faulty, that humans are not rational – effectively, that humans are not humans. This is not to say that human beings, being rational by definition, as it were, are infallible. The main point is that the meaning of 'rational' is not arbitrary but discoverable via an investigation of what humans say and do. If we work with Hume's premise, we are abusing language by violating the implicit criteria in the use of words like 'reasonable', 'rationally acceptable', 'justified'. Moreover, to understand what we mean by 'induction' and discuss its alleged problems, we must always accept its authority in the very working of the language we are using.

There is, therefore, a kind malignant of reductionism involved in Hume's celebrated attack on induction. Its dangerous character is evident in the fact that it involves some very fundamental philosophical problems dealing with the very meaning of terms. The diagnosis of the problem of induction as I sketched it up to now illustrates why arguments and counter-arguments regarding this area are essential for my original aim. As I mentioned at the beginning of this section, my aim is not to present an overall survey of all arguments related to induction but to concentrate on those that highlight the necessity of involving the acting person within the discussion. What we are dealing with here is the ordinary-language approach to the problem of induction. Further inquiry is in order, especially to bring out the inevitable link between the methods of science and the speech-acts of the individual scientist.

Let me start with the general observation that philosophers in the ordinary language approach often argue that it does not make sense to ask for a justification of induction. Two reasons are frequently given here. The first is that it is *analytic* that a valid inductive inference is an inference of such-and-such a kind. In other words, the term 'valid' just means 'inference of the inductive kind'. This seems to be the kind of reason at the back of, say, Ludwig Wittgenstein's comments on induction in the *Philosophical Investigations*.[10] The second reason is that there are no standards by reference to which inductive standards could be justified. In other words, induction is a kind of rock-bottom experience, so much intertwined with our everyday, rudimentary action that its status is supreme. It sets the standards of validity.

Even though these two reasons weaken the case for an effort to justify induction, they do not undermine it completely. There is still the question whether the concepts we use should indeed be defined as they actually are. In other words, even if 'induction is valid inference' is analytic, there is always the need to examine whether our concepts should be such that it is so. It is not obvious that the meaning of 'valid' should be linked exclusively to 'inductive inference'. It is not obvious that the ordinary use of these words should be accepted as normative without first being

10 For instance *Philosophical Investigations* §§ 324–325. The current debate on the ordinary-language approach took definite shape with P. Edwards' article: 'Russell's Doubts about Induction', *Mind*, 68 (1949): 141–63.

ratified by philosophical argument. Hume has famously discussed the disanalogies between valid cases of deductive reasoning and valid cases of inductive reasoning. An important question, however, is often overlooked by commentators of Hume's approach: what are the analogies between cases of deductive reasoning and those of inductive reasoning that justify the use of the same word 'valid' in both cases?

L.J. Cohen, in his book *The Implications of Induction*, starts his discussion of this point by recalling that the various arguments presented to defend induction are either of the form of a validation or of the form of vindication.[11] Validation means showing that the inductive argument as a logical form is valid because, although it uses special principles, such as the principle of the uniformity of nature, these principles are all independently accredited. Vindication means showing that it is a good policy to adopt inductive inquiry as a heuristic method, because, for instance, it is self-correcting. Validation, therefore is a matter of good logical form, while vindication is a matter of good policy.

Now the justification of induction is often presented as an extremely important task, given the ubiquity of this mode of reasoning in science and in everyday life. If induction is not justified, the entire edifice of our intellectual endeavours will apparently be in serious danger of caving in. What I will call the ordinary language argument, however, illustrates that this alarming claim is artificial and hollow. Requests for the justification of induction are trivial.

The ordinary language argument is the following. In a given situation that requires us to arrive at a conclusion, there is simply no space between the idea of being reasonable and the idea of arriving at a conclusion by adjusting our degree of conviction according to the strength of inductive evidence, understood here as the number and variety of observed cases. In such a context, 'being reasonable' just *means* engaging in an inductive procedure. It is therefore pointless to ask for a proof that one is reasonable when one relies on induction.

A variation of this argument can also be expressed in terms of standards, understood here as evaluative criteria. When we ask whether the use of induction is justified, we are assuming that there are standards of justification we may appeal to other than the standards constituted by induction itself. This assumption, however, is wrong. It goes against the obvious fact that inductive procedures are a fundamental ingredient of all our everyday thoughts and actions. There are no standards better grounded than induction itself. It is therefore senseless to ask the question in the first place.

An argument may be put forward to suggest that this version of the ordinary language defence of induction has a weak point. One starts with the idea that analyticity is a semantic attribute that needs investigating. Non-evaluative terms may be analytic but evaluative terms are not.[12] Take for instance the non-evaluative term 'solid'. This has a meaning that is derived from paradigm cases in ordinary

11 L.J. Cohen, *The Implications of Induction* (London: Methuen, 1970).

12 Cf. J.O. Urmson, 'Some Questions concerning Validity', in A. Flew (ed.), *Essays in Conceptual Analysis* (London: MacMillan, 1956), pp. 120–33.

language. There is analyticity between 'solid' and 'of the consistency of such things as rocks'. For evaluative terms, such straightforward equivalence of meaning is often impossible. G.E. Moore's famous argument about the term 'good' is a case in point. There is always some space between 'good' and any other expression we care to link it up with. It is always possible to add the further evaluative question: 'Is *this* good?' The same thing seems to be the case as regards 'valid'. This is also an evaluative term. There is apparently the space to ask the further question: 'Is *this* valid?' If this is correct, the weak point of the ordinary language defence of induction would lie, therefore, within the very idea of analyticity that constitutes its foundation stone.

The proper way to reply to this objection is to uncover the artificiality of the alleged resemblance between 'good' and 'valid'. The former term is associated with an enormously broad spectrum of things, events, and persons that may or may not be the bearers of goodness. The latter term, on the contrary, is an attribute only of one kind of thing, namely inference. Moreover, one cannot escape the simple fact that ordinary inductive standards are the only standards there are to decide about validity. There is a certain set way how the word 'valid' is used. We either accept this and be accepted within the linguistic community, or we opt out altogether.

A more serious objection to the ordinary language defence of induction can be formulated if the very idea of ordinary language is put into question. Cohen expresses this point as follows: 'in addition to discussing how terms are actually used we can also discuss the merits of those usages' (p. 187). He draws inspiration from advances in mathematics that involved the introduction of new uses of already established terms. For instance, the inclusion of zero as a number, the inclusion of the square root of minus one as a number, and the inclusion of Lobachevsky's postulates as geometry are all cases of an extension of an already established use of some central terms, in this case 'number' and 'geometry'. The challenge is therefore to see whether the term 'number' is *justifiably* used in a sense that makes it true to call zero a number. Mathematicians have answered this question in the affirmative. If we shift our attention now to the case of induction, instead of discussing the merits of the extension in the use of the term 'number' or of the term 'geometry', we discuss the merits of the extension in the use of 'valid'.

The objection to the ordinary language defence of induction is essentially the claim that some extension of the term 'valid' is not justified. Consider the use of the term 'valid' in the case of reasoning from premises to logically implied conclusions, as one does in syllogistic reasoning. If we follow the lead of Hume on this point, this use may be taken as the paradigmatic, or well-established use of the term 'valid'. We proceed by extending this use to cover also cases of reasoning from premises about the already observed to conclusions about the yet unobserved. These cases are those of induction. The question then arises whether the extension in the use of 'valid' is justifiable or not. Hume has famously insisted on the pervasive lack of analogy between the syllogistic-use of 'valid' and the inductive-use of 'valid'. Such lack of analogy convinced him that the term 'valid' is inadequate to support such extension.

Various counter-attacks have been mounted against this objection. It is helpful to consider such attempts as falling under two kinds. To counter such an objection, one can attempt to *validate* induction, or one can attempt to *vindicate* it. In general, validation is a matter of epistemology. One validates a claim to knowledge. Vindication, on the contrary, is a matter of practical reasoning: one vindicates an action.[13] As regards induction, validation takes the form of presenting inductive reasoning as relying solely on the logical or mathematical criteria of valid reasoning, understood in terms of deduction. This is essentially a logical exercise meant to construct a bridge between induction and deductive reasonableness. It is essentially an exercise of 'deductifying' induction. As opposed to this, vindication takes the form of a search for a set of criteria to show the reasonableness of performing certain actions. In this way, the vindication of induction is intended to show not its deductive reasonableness but its practical reasonableness.

Both these moves remain somewhat unsatisfactory principally because they do not consider induction as a *sui generis* way of inferring. They seek instead to show that it is acceptable in so far as it is allegedly a special case of non-inductive inference. To take induction seriously, as a *sui generis* way of inferring, one needs to respect its undeniable role in everyday life and thence examine what makes it so central. This means, essentially, showing the legitimacy of the everyday use of such expressions as 'is validly reasoned from' when used as regards both deductive and inductive procedures. There is no doubt that in everyday life there is a high degree of analogy between induction and deduction, neglected by Hume. In fact, where Hume went wrong was in his neglect of elements of similarity, elements of positive analogy, between the two uses of the term 'valid'. His reasoning is correct as far as the distinctive character of each way of inferring is concerned, but misguided in its selective treatment of the issue. What must be done, therefore, is to complete Hume's inquiry into the comparison the use of 'valid' as regards the two ways of arriving at a conclusion. Completing this task means here retrieving the analogies that he ignored, so as to have a more responsible picture and be able to judge whether the analogies carry more weight than the disanalogies, or *vice versa*.[14]

This move is called the *ratification* of induction by Cohen because it deals with showing the legitimacy of its everyday use as implicitly agreed upon by a group of people. The basic insight here is that the basic source of validity lies in the smooth running of life in general. The term 'valid' is attributed to inferential procedures when they allow, and even uphold, the smooth running of life. Such attribution is done both in cases where the conclusion is logically true and in cases where the

13 The distinction comes from Herbert Feigl, 'De principiis non disputandum...? On the Meaning and the Limits of Justification', in M. Black (ed.), *Philosophical Analysis: A Collection of Essays* (Ithaca, NY: Cornell University Press, 1950), pp. 119–56.

14 This kind of inquiry finds a interesting precursor in Aristotle's famous treatment of different kinds of good (*Nicomachean Ethics* Book I, chapter 6) where he assumes that the use of 'good' for various different cases is not senseless but evidence of some significant analogy.

conclusion has such-and-such inductive support. The significant analogy between deductive and inductive inference, which Hume neglects, derives from the way everyday life-situations proceed with success, most of the time. The inference from A to B's being logically true and the inference from C to D's having such-and-such inductive support are both considered valid, by popular consent as it were, because they both allow and uphold the smooth running of everyday situations. Admittedly, there may be times when induction leads to a mistake and therefore to a bump, as it were, in the smooth running of life. This fact, however, should not make us forget that the smooth running of life would also be disrupted, this time much more seriously, if all inductive inferences were considered untrustworthy.

The question was to see whether the analogies between induction and deduction carry more weight than the disanalogies. A clear reply now is possible because of the asymmetry that has become evident between, on the one hand, the very seldom bump that arises in the smooth running of life when both inductive and deductive inferences are considered valid, and, on the other hand, the complete disruption of life when all inductive inferences are considered suspicious. This asymmetry shows that the analogies are more important, as regards everyday life, than the disanalogies. It is consequently an indication that there are good reasons for the use of 'valid' as regards inductive inference, as consensus demands. One may worry that some disanalogies between induction and deduction are ineliminable. As long as they still lurk in the background, the problem of induction remains. Such a worry, however, is a fossil of the desire to deductify induction. Such a project is better considered extinct.

This vindication of induction offers us renewed confidence in induction as a valid method of inference supporting the method of science. It suggests that by the process of induction we are in fact not trying to invent some kind of regularity between two unrelated features of the world, but trying to figure out, or converge upon, *the existing regularities inscribed in the world independently of us*. Hence, when we use induction to arrive at a conclusion from observing a finite number of swans, we are not inventing a link between all the observations. We are making a reasonable guess at the regularity that exists independently of us, a regularity in the nature of these objects. It does not matter much that we may go wrong, either in our understanding of what swans are, or in our understanding of what the specific regular features are, the whiteness in this case. The point is that induction is a heuristic mode of thinking directed towards the identification of already existing regularity.

Securing this confidence, however, was not my primary aim. I analysed the ordinary-language approach precisely because it brings out the essential link between basic scientific methods and the everyday life-situations of the individual. This approach makes it clear that inductive logic is not unrelated to how individuals conduct their lives. The general aim in this book is to say something about the way the scientific mentality may have some influence on the personal attributes of the individual scientist. Inductive logic forms part of this mentality and is related directly to everyday life. It is highly recommendable therefore to engage in a study of how induction may influence the personal attributes of the individual. I will leave such a study to the last section of this chapter, after some evaluation of the other pole of standard scientific methodology, namely falsification.

Falsification

Is the link between purely logical discourse and what people typically do in everyday situations as important for the correct understanding of falsification as it is for induction? In this section, I will argue that the answer is yes. My overall project, let it be recalled, is to explore how the scientific mentality affects moral character. Showing that falsification is intimately related to everyday life is a precondition for such a project.

The best way how to start an inquiry about the nature of falsification is perhaps to recall that scientists necessarily engage in their observation-activity already marked with certain presuppositions. These presuppositions are normally structured together in a kind of hypothesis offering a first tentative description of whatever they are studying. One may describe therefore the role of the scientist as one of testing a hypothesis to see whether it clashes with observations. In this way, one will be giving an account of science without the need of the inductive method. The scientist tests a given hypothesis by hunting around for a possible observation that will clash with the hypothesis. If such an observation is found, the scientist will discard the hypothesis, and take up the next hypothesis on the list, as it were. Such an account of science is based on the basic idea of falsification of conjectures, an expression introduced in this context by Karl Popper. It is an account that enables competing hypotheses to be viewed as analogous to biological species whose individual organisms are struggling for survival. What determines which one species survives is the combination of characteristics of the species and aspects of the environment. In a similar way, what determines which hypothesis survives is the combination of various characteristics, including the content of the hypotheses themselves, the scientist's skill at coming up with appropriate tests, and also the actual world the competing hypotheses are meant to describe. The community of scientists here is viewed as an open society in which everyone has the right to propose new conjectures and everyone has the right to criticise. Everyone seeks the truth. But everyone is aware of limits and uncertainties.

It may be useful here to highlight the two main differences between this new account of science and the one discussed previously, the one based on induction. The differences are best appreciated as regards the task of the scientist. First, the scientist within the inductive picture is someone whose job is to prove that a given theory is true. Such an operation is called verification. As opposed to this, the scientist within the new picture is someone whose job is to refute a given hypothesis. Such an operation can conveniently be called falsification. The second difference has to do with the role played by the scientist's presuppositions. Within the verification picture, scientists are encouraged to free themselves of any presuppositions as far as possible, so as to arrive at their observations free of bias. Complete freedom from bias, as we have seen, may not be realisable. But the point here is that the role of the inevitable presuppositions should be minimised. The exact opposite happens within the falsification picture of science. Here, scientists are encouraged to be as imaginative and audacious as possible. They are encouraged to come up with interesting new conjectures that fit the observations in ways that are more and more acceptable. Moreover, this non-inductive account of science can distinguish between authentic and inauthentic scientific proposals. If natural science is essentially the falsification of conjectures, then any discourse which in principle cannot be falsified by any conceivable experiment can never be the object of study of the natural sciences. Hence, if one accepts this account of science, one has a way of demarcating between the discourse of natural science and other kinds of discourse. This distinction does not say anything about whether a given proposal is actually false or not. It is saying something about a possibility – which we may call the *falsifiability* of the proposal.[15]

A historical example may be helpful to see what these changes imply in the real life of scientists. A genuine scientific attitude is often considered to be one that makes the proponent of a new theory ready to abandon his or her proposal whenever a falsifying observation is made. This genuine scientific attitude can be seen in Albert Einstein's readiness to abandon his General Theory of Relativity if contrary evidence were to be found. One particular conclusion that can be derived from this theory is that light does not always travel in straight lines, but will follow a curved path when passing close to a heavy astronomical body. On this issue, the entire theory could have been falsified. Historians tell us that Einstein was ready to abandon his theory if light rays could be shown not to be curved near a massive object like the Sun. The crucial test was carried out by Arthur Eddington during a total solar eclipse of the Sun in 1919. A star whose light should have been blocked by the eclipsed Sun could nevertheless be observed. This was precisely because the star's light was deviated slightly as it passed near the Sun. It could thus travel 'around' it. The displacement of the star's position, corresponding to the curvature of the light's path, was of the same magnitude as that predicted by Einstein's General Theory. The theory was thus not falsified on this crucial issue. In fact, it gained considerable credibility.

In a sense, this example from history recounts a success story. Not all episodes in the history of science are like it. What I have called the genuine scientific attitude is unfortunately not always present. Some researchers may present a conjecture to the scientific community but will not be ready to abandon it if falsifying evidence becomes available. Defenders of Freudian psychoanalysis and Marxism have been accused of lacking the genuine scientific attitude in this way.[16] Not to be dragged

15 One should recall here that this distinction is not the one suggested by the Vienna Circle: it is not the distinction between meaningful discourse and meaningless discourse. The distinction here is between the discourse of natural science and other kinds of discourse. These belong to other categories: poetry, mathematics, metaphysics, religion, and so on.

16 The accusation has been made most explicitly by Karl Popper. See also his related discussion on the demarcation between science and non-science in: *The Logic of Scientific*

into a complex debate about the merits of psychoanalysis and Marxist social theory, I would like to keep my comments on this point to a minimum. It is enough here to highlight the fundamental point: without the genuine scientific attitude, the importance of contrary evidence is disregarded. Intellectual honesty is thus jeopardised because the problems are swept under the carpet of our presuppositions.

A basic question has been systematically avoided up to now: with this non-inductive account of science, can we say anything about the truth? It seems that what we are entitled to say is limited to two things. We are entitled to say whether given proposals are scientific or not. We are also entitled to say whether given proposals have been falsified or not. On this view, the scientific judgement seems to stop short of any claims about what the world is actually like. Science is interested in theories and not in the world. Is this a real limitation of our non-inductive account of science?

To answer this question, one needs to make a distinction between two kinds of judgement. Given a hypothesis, one needs to distinguish between, on the one hand, how much it merits our acceptance, and, on the other hand, how much it faithfully describes that part of the world it was meant to describe. The two kinds of judgement are not the same. We may have some criteria to decide about the first, but none about the second. It is indeed difficult to conceive of relevant criteria for the second kind of judgement. In other words, it seems difficult to determine criteria to help us decide how faithfully a theory describes reality. The problem here is that we apparently cannot have an idea of the material world other than through a theory. It seems that what we can compare is one theory with another theory, and not one theory with the world.

Let us therefore explore each one of these two kinds of judgement. The first one concerns how much a proposed theory merits our acceptance. Presumably, a theory can merit our acceptance a lot, or a little. To describe this aspect, a useful idea to use is that of 'corroboration'.[17] A theory is well corroborated when it is not only easily testable but also strong enough to pass severe tests. A theory is easily falsified when it is simple, bold and very precise, like when we say 'All *A*s are *B*s'. A theory that is not easily falsified, of the form 'Some *A*s are *B*s', cannot merit our acceptance to the same degree as one which is. The more severe the tests that are passed by a theory, the more it merits our acceptance. A test is severe when it involves a surprising consequence of the proposed theory. In other words, a severe test usually concerns a prediction of the theory that is very improbable.

The bending of light is a very good example of a highly improbable prediction of the General Theory of Relativity. The fact that the theory has passed this test says a lot about how well corroborated the theory is. It says that, from that moment onwards, we should accept it strongly. It may not have been very acceptable before the crucial observation of the bending of light. After that particular observation, however, it is well corroborated. This example shows that whether a theory is well-

---

Discovery (London: Routledge, 1992), chapter 1, § 6.

17 A key term for Popper, see Ibid., chapter 10.

corroborated or not depends essentially on the tests that the theory has actually passed. The theory is similar to the runner who holds the 100-metre world record. The runner starts to deserve our considering her a champion only after the crucial race during which she actually breaks the previous record. And she will continue to deserve our honouring her with the title until a faster runner will come forward and break the existing record. Well-corroborated theories are those that have done very well up to now, according to the tests scientists have been able to devise. New instruments will inspire new tests. And these developments may change the degree of corroboration of theories. Olympic judges cannot tell whether a runner is a record breaker or not if the runner has never run a race. They naturally tend to think that only those who do not participate in the race do so because they are not good enough. Similarly, if no tests are available for a particular theory, scientists hold that it isn't well corroborated at all. If scientists had systematically missed all total eclipses of the Sun since the proposal of the General Theory of Relativity, the theory would not have gained any degree of corroboration – at least in the area concerning the bending of light.

This is therefore one kind of judgement. The main point here is to see whether a proposed theory deserves our acceptance or not, and this decision depends on the theory's degree of corroboration. The object of what I called the second kind of judgement involves another question, one that is more important, namely: is the theory true? This is a different aspect of a proposed theory. The truth-value of any descriptive statement does not depend on the particular phase in the history of science when it was uttered. In a sense, its truth-value is a-temporal. Whether we know that it is true or false, is another matter.

When we extend these reflections from the realm of individual statements to that of entire scientific theories, we see that the same thing may be said, be it with some slight modification. A theory may be roughly described as a vast number of statements linked together in certain logical relations. Some of these statements will be logical consequences of the entire system, and some of these logical consequences will be true. Hence theories, understood in this way, may be said to show a certain degree of truth-likeness. Truth-likeness, or verisimilitude, has to do with how good an approximation the theory is.[18] It depends on how many logical consequences of a given theory are true. If a theory has a great number of true logical consequences, then it has a high degree of truth-likeness. If a theory has a small number of true logical consequences, then it has a low degree of truth-likeness. The degree of truth-likeness proposed here is quite different from the degree of corroboration the

18 The idea of verisimilitude is introduced and discussed by Popper in: *The Logic of Scientific Discovery*, Appendix IX; more extensively in *Conjectures and Refutations: the Growth of Scientific Knowledge*, 3rd ed. revised and enlarged (London: Routledge & Kegan Paul, 1969). For a valuable exposition of Popper's views see C. Britz, and C. Brink, 'Computing verisimilitude', *Notre Dame Journal of Formal Logic*, 36 (1995): 30–43; I. Niiniluoto, 'Verisimilitude: the third period', *British Journal for the Philosophy of Science*, 49 (1998): 1–29; P. Tichý, 'On Popper's definition of Verisimilitude', *British Journal for the Philosophy of Science*, 25 (1974): 155–60.

theory enjoys at a particular period in history. Just like the truth or falsity of a single statement, the degree of truth-likeness is independent of whether we know anything about it or not. It is independent of our experimental techniques available to test the theory. Whether a given theory is a good approximation or not is not tied down to any particular period in history. It is a timeless property of the theory.

Now that I have distinguished between corroboration and truth-likeness, I have the necessary tools to tackle my previous question: with a non-inductive account of science, can one say anything about the truth? The answer to this question lies in the relationship between corroboration and truth-likeness. Given a new theory, scientists will endeavour to construct severe tests so as to falsify it. These endeavours will obviously be dependent on the techniques available. Through this work, scientists will arrive at an idea of how well corroborated the theory is. Now comes the crucial jump in their reasoning: scientists will argue that the only way they can get an idea of how good the theory is as a description of reality is precisely through its performance as it confronts the severe tests they devise. Hence, the only way of guessing the degree of truth-likeness of a new theory is to notice how well corroborated it is. In other words, a very well-corroborated theory is a very good approximation to the truth. Otherwise, it would have fallen off by some test or other.

This way of handling the question of truth seems fairly plausible. It apparently shows that the non-inductive account of science can indeed offer what it had promised, namely a coherent, new way of understanding scientific work that is entirely deductive. In spite of this *prima facie* plausibility, however, there are at least two points of the proposed deductive account through which induction still leaks in, as it were. My aim is not simply to identify the inadequacy of a purely deductive system. It is rather to highlight the fact that induction becomes inevitable precisely at those points where falsification, as part of the scientific method, is connected to the everyday practice of normal human beings.

The two weak points of the deductive structure start becoming evident when Popper takes the crucial step of linking well corroborated theories with verisimilitude. His struggle with the question of truth is obvious in the following paragraph:

> Science is not a system of certain, or well established, statements; nor is it a system which steadily advances towards a state of finality. Our science is not knowledge: it can never claim to have attained truth, or even a substitute for it, such as probability. Yet, science has more than mere biological survival. It is not only a useful instrument. Although it can attain neither truth nor probability, the striving for knowledge and the search for truth are still the strongest motives of scientific discovery.[19]

It is clear here that Popper is not making the simple claim that survival of severe tests helps to verify theories. He is bestowing, rather, an important role to the 'search for truth' as a driving force. It is highly natural for us to accept that a well corroborated theory is a reasonably good picture of the world. Such a claim exposes our realist attitude, in the sense that we hold that there is a world that exits independently of our

19 Popper, *Logic of Scientific Discovery*, p. 278.

ideas about it, and that this world is by and large more similar to the way modern theories describe it than to the way superseded theories describe it. How could a theory be well-corroborated if there isn't 'some truth' in it?

This question sets the scene for the first point in Popper's proposal where induction leaks in. This first point is acknowledged by Popper himself. At one point in his 'Replies to my Critics', he makes an effort to show that arguing about scientific progress in terms of verisimilitude is better than arguing in terms of truth.[20] He concedes the import of the intuitive claim that a theory is well corroborated because there is some truth in it. There are two ways of understanding this loose way of expressing our intuition. One way is to limit oneself to a double negative, that is to hold that it is improbable that the theory is not true. From here, we deduce that it is probable that the theory is true. This way employs the notion of truth, and is deductive. The other way employs the notion of verisimilitude. It involves interpreting the intuitive statement as saying that it is probable that the theory has a higher degree of verisimilitude than those of its competitors. This second way seems to be non-inductive. To see why this is so, consider how theories are judged as regards their degree of verisimilitude. In Popper's system, if a theory is initially considered highly acceptable and then survives various tests, the probability that it has a high degree of verisimilitude is low. The reason for this is that, in the testing procedure, there is no element of surprise. It is precisely this element of surprise that is needed for the probability we are seeking. The opposite kind of theory is one that is initially considered not probable at all and then eventually survives various tests. In this case, we are entitled to say that it is probable that this theory has a high degree of verisimilitude. In other words, the probability that a theory has a high degree of verisimilitude is inversely proportional to the initial probability of the theory, prior to testing. This inverse proportion is what convinces Popper that induction is not involved here.

More careful analysis, however, shows that, in spite of this inverse proportionality, induction is still present here. Popper himself makes the following concession:

> In spite of this [inverse proportionality], there may be a 'whiff' of inductivism here. It enters with the vague realist assumption that reality, though unknown, is in some respects similar to what science tells us or, in other words, with the assumption that science can progress towards greater verisimilitude.[21]

The best way to interpret this seems to be the following. Popper is here acknowledging the argumentative weight of the simple kind of realism without which everyday life would be impossible. What he calls 'the vague realist assumption' connects what happens in everyday life to what happens in the evaluation of complex scientific

---

20 Popper, 'Replies to my Critics' in P.A. Schilpp (ed.), *The Philosophy of Karl Popper* (Illinois: Open Court, 1974), pp. 961–1197. The discussion is situated within a reply to A.J. Ayer's accusation that Popper's idea of verisimilitude is worthless. There is also a significant long footnote; pp. 1100–3.

21 Ibid., p. 1193.

theories. In the former situation, we are simply obliged to manage our lives, as we grow from infancy to childhood and beyond, by assuming that, by and large, our understanding is not an empty exercise but is an ever-increasing understanding of the real world. Such an assumption, in fact, is better called a principle, because there is no definite point in time when we consciously choose to endorse it. For basic appropriation of language, it is a logical necessity. In the situation involving the evaluation of scientific theories, Popper is acknowledging that the same principle has still some role to play. As has been stressed by various commentators, saying that science progresses towards greater verisimilitude, as Popper does, amounts to an inductive argument.[22] The additional point I would like to highlight here is that Popper's 'whiff' of induction enters precisely through the link between scientific theorising and the 'vague realist assumption' of everyday life situations.

Having secured this link between falsificationism and everyday practice as regards the first weak point of Popper's deductive structure, I move on now to the second. This concerns the search for counter-evidence: the search that forms one of the main tasks of the Popperian scientist. Can this particular task be carried out non-inductively? The answer is no. Consider a simple example. We start with a conjecture in the form of a general law, say 'All swans are white'. We then wait for scientists to come up with a counter-example. The counter-example is expressed in a statement of this form: 'There is at least one swan here that isn't white.' This statement by scientists seems to be a simple instance of observation – it concerns only one swan. One must recall, however, that the observation of this one individual counter-example must have been made not just at one isolated moment, but repeatedly. It is an illusion to think that scientists can content themselves with reporting a fleeting instant. Fleeting instants do not exist. In real life, the swan that will knock down the original conjecture, let's say a black one, must be a swan that is black now, was black an hour ago and will, for all we can say, be black an hour hence. It is this kind of swan that is reported as a non-white swan, and thereby a significant counter-example. The claim that something is non-white is an empirical observation. It therefore necessarily involves a stretch of time, a duration. To have a counter-observation, scientists will certainly need to indulge in an inductive method: they must infer, albeit unconsciously, that at moment m_1, this swan is black, and at another moment m_2, the same swan is black, and at another moment m_3 the swan is black, and so on, the conclusion being that this swan shows all the characteristics that normally make us affirm that it is black. This is how the observation, as an event within the life we know stretched out along the temporal axis, attains the power to knock down the general conjecture. In the last analysis, therefore, observation involves plunging oneself into the temporally extended stream of life. Even though the logical structure of the falsificationist method, at least in its non-sophisticated form presented by Popper, is neatly deductive, the process of supplying the element of counter-evidence, which is one of its essential ingredients, involves some elements of induction. I make this claim even though I also admit that

22 For instance, Anthony O'Hear, *Karl Popper: the argument of the Philosophers* (London: Routledge, 1980), Chapter IV.

there is a level of reflection that goes beyond induction. As I explained in Chapter 2, this level involves the analysis of meaning, so as to see how concepts are logically, as opposed to empirically, dependent on experience. At this basic level, our use of words is non-inductive. The experiments involved in science, however, are far from such foundational speech-acts. Even in the mere observation of a counter-example, the inductive element becomes significant. Here again, therefore, there are enough reasons to support my claim that falsification is open to some inductive procedures precisely because of its inevitable link to what happens in the everyday life of the individual scientist.

To sum up, a full understanding of falsification just like that of induction discussed in the previous section, cannot neglect the essential link between, on the one hand, the purely logical discourse and, on the other hand, what people typically do in everyday situations. My argument started with an illustration of Popper's basic mechanism, namely the one whereby the only way of guessing the degree of truth-likeness of a theory is to notice how well corroborated it is. I proceeded then to identify two areas in this deductive system where induction seeps in. The upshot is that induction becomes inevitable for Popper precisely at those points where falsification, as part of the scientific method, is connected to the everyday practice of normal human beings.

The residue of inductive elements within the deductive system of falsification should not be taken to imply that nothing has been gained by Popper's reasoning. In fact, it is undeniable that, within the task of scientists, a fundamental role is played by the falsification of mistaken hypotheses. It seems therefore that a satisfactory account of science must include the good points of both this non-inductive approach and also the inductive approach, which in some form or other is inevitable anyway. Sophisticated forms of inductivism and falsificationism need to be determined, as will be discussed later.

Method and Virtue

Having discussed at some length some aspects of the scientific method, I have enough material now to explore the effects this method might have on the personal attributes of the individual engaging in it. The basic insight that allows such an exploration is that the method discussed is essentially connected to everyday life. Any clear breech between method and life can only be artificial and therefore misleading. I take this point to justify my claim that methodology is relevant for an inquiry about the way persons live. A discussion of virtues, as was indicated in the previous chapters, is primarily concerned with the transformation of those attitudes or dispositions that have a bearing on conduct and decision making, on the orientation of a person's life in general, and ultimately on whether the person is a good person or not. Of the points presented in this chapter, two appear more important in this regard. The first concerns the deeper aspiration of science to approach as much as possible a kind of knowledge that is necessary and universal. The second concerns the entire issue of how to legitimate scientific claims, either through verification or falsification. The procedure will be similar to that undertaken in the previous chapter: I will first examine each of these points so as to arrive at a pair of opposing vices associated with it. For each point then, I will determine, in so far as it will be possible, the particular virtue situated between these two vices.

As regards the first point, it goes without saying that some kind of aspiration to arrive at universal judgements forms part of everyday life. Making sense of what is immediately encountered around us is itself a move towards forming judgements that, in their broad application, reach out beyond the specific cases one deals with. This basic attitude therefore illustrates how the very act of living constrains us to assume that the world has a regularity worth discovering. One needs to trust the world, at least to a certain extent. One acts on the hope that some judgements latch on to the world's regularities in such a way that they be valid not only from one particular point of view but from all points of views. The scientist's aspiration to arrive at judgements that are necessary and universal is, therefore, not something distinct from common attitudes and common sense. It is, in fact, a continuation of what one does in the modest setting of everyday life, where one often seeks a platform of knowledge that is valid for all people. Seeking such a platform is desirable because it is useful. Agreement and co-operation, and even the very existence of language, depend on it. The danger is that, when one only aims at gaining such a common platform of knowledge, one easily slides into a neglect of the legitimate differences between individuals. Yielding to such a temptation results in yearning for a simple monolithic universe in which only what is universal counts. Assertions that are bound to a specific point of view will be considered of marginal importance or even empty or unreal. In other words, being too keen on arriving at what is universal and necessary may make a person cling to a simplified view of what deserves attention. It may make him or her overestimate the importance of claims of universal validity. Cultural differences become an unfortunate aberration, a human weakness to be conquered. This is an extreme attitude. When ingrained into the life-style of an individual, it results in the habit of levelling off all differences, lawn-mowing the global, social landscape in an act of cultural totalitarianism. In this extreme form, it is obviously a vice to be avoided.

The opposite vice can be seen to result from an awareness of an important aspect that has not been highlighted so far. The search for what is universal is in fact a move away from the individual. A gap becomes evident between, on the one hand, knowledge claims in the form of universal and necessary statements, and, on the other hand, assertions, together with cultural and social aspects that depend on personal taste and opinion. The individual scientist in this case will typically be led to live a double life. In the laboratory, he or she will be living in a world totally dissociated from the normal everyday world outside the laboratory, from the world of opinion, of likes and dislikes. In the extreme, this tendency may lead to a situation in which a single individual gets into the habit of switching from one world to another without ever attempting to bridge the gap between the two. The vice associated with this trend may be described as a fragmentation of the self, or a lack of personal wholeness.

When this vice is taken together with the opposite vice described previously, the virtue associated with the particular aspect of scientific mentality discussed here can be determined. The wise person will try to trace a path midway between the excess of aggressively levelling off all differences in the name of the exclusive search for universal knowledge, and the opposite excess of not caring anything at all about such differences, in the name of a passive acceptance of social and even personal fragmentation, leading, on the broader scale, to a kind of cultural apartheid. As has already been explained in the first chapter, virtues are essentially habits that are good in the sense of producing acts conducive to promoting human moral good. In this case, the typical scientific method of seeking universal and necessary knowledge can be seen as possibly enhancing the virtue of reflective equilibrium, whereby the person is balanced, conciliatory, and non-extremist. I intend the term 'reflective equilibrium' to refer to the virtue manifested by acts that give due importance both to what is universal and to what is particular, especially as regards the maintenance of harmony between peoples that show considerable cultural variety and creativity.

This virtue concerns the first point mentioned above, namely the aspiration of scientists to gain knowledge that is necessary and universal. The second point that deserves special attention as regards possible effects on the individual's attitudes is the one dealing with the logical structure of the method involved. Induction and falsification have been presented as heuristic strategies that are indispensable for scientific work. To determine the virtue that this aspect of the scientific mentality could enhance, given the right conditions, I will proceed in the same way as before. The first task is to determine the two opposing vices associated with this aspect. This done, I will describe the virtue by considering the mean between them.

As regards induction and falsification, the basic question seems to be one of trusting our intellectual abilities. To what extent can we trust our normal ways of understanding as regards what we desire: presumably a correct version of what is the case? The previous discussion on induction exposed at least one worrying attitude that may arise in this domain. It was shown that drawing a conclusion by induction is never as secure as drawing one by syllogistic deduction. Overgenerous use of induction may easily result in an excess of false claims about the world. With the good intention of trusting our intellect, we may eventually end up having an unfortunate disposition of seeing more order in nature than there actually is. This may become ingrained as a vice in our personality. I am referring here to the bad habit of wanting to reduce nature to a clean, simplistic model. This is equivalent to the habit of avoiding giving due importance to exceptions or irregularities. Such a habit explains how some people tend to content themselves with the physics of the ideal case. The effects of this vice have been the subject of much derision. Special targets in the course of history have been those thinkers who engaged in the production of theories about what constitutes the deep nature of reality on the sole constraint that their account be self-consistent, or consistent with traditional world-views that have only poetic value.

We are dealing here with a vice that corresponds to what was poetically called the 'idol of the tribe' and the 'idol of the square' by Francis Bacon. His intention in using this terminology was to encourage his readers to renounce or destroy such idols and seek the truth. If my argument is sound, however, induction alone cannot do the job Bacon desires. It has aspects that can generate the opposite vice; it can produce, as it were, another idol that Bacon never thought of. I have highlighted the fact that critics of inductive reasoning typically express worries about the rationality of anyone employing it. Hume's famous appeal to the idea of habit to serve as an explanation does little to redeem its status. On being exposed to such arguments, scientists realise that induction, although all-pervasive, seems to be of dubious value. They can therefore easily slide into a sceptical attitude as regards all acts of understanding. Any kind of claim starts being considered guilty until proved innocent. Heuristic risk-taking will be considered outlandish, even irrational.

This kind of personal disposition is not limited to those engaged in inductive reasoning. It is also present in some form or other in the lives of individuals who limit themselves to the method of falsification. This happens because the method of falsification was proposed as a way of avoiding the alleged inadequacy of induction. People limiting themselves to falsifying conjectures can easily slide into the habit of forgetting their fundamental aspiration to latch on to the external world. Their main interest is concentrated on the proposed hypotheses rather than on the world itself. Taken to the extreme, this habit may make the individual seek fame exclusively through the destruction of the work of others. The individual thus tends to become an expert at throwing stones at the edifice of received views. In other words, he or she tends to become a master-destroyer rather than a master-builder. The very idea of exploration and discovery is undermined. The person becomes over-cautious and ultimately incapable of moving forward, somewhat like the child who never learns how to ride a bike because of over-concern with what can go wrong while moving the legs to push the pedals, with what can go wrong while tilting the handle bar, with what can go wrong while leaning over to go round corners, and so on. As in other cases discussed previously, the vice that is being described here should not be considered a necessary result of repeated use of the method of falsification. My claim is rather that it is a possibility with high probability. Such a possibility can be conceived when the life of the person is considered in the long run, exclusively engrossed in that particular method. As such, this negative habit is therefore a limiting case, perhaps rarely occurring in reality.

It is not difficult to see that the two vices described so far are opposing vices. The first one involved being so easy with inductive reasoning that too much regularity is imposed onto the description of nature. The second involved being so engrossed with the falsification of proposed conjectures that our contact with nature withers away. The first is therefore an over-confidence as regards our mental ability to arrive at correct understanding of the world; the second, an underestimation. The virtue to be sought therefore is the mean between these two extremes. It involves having the right kind of confidence as regards our intellectual abilities, neither forgetting the possibility of making mistakes nor being overcome with the fear of making them to the extent of wanting to avoid all risks whatsoever. It corresponds to a form of prudential risk-taking, together with a correct appreciation of both the strong points and week points of our intellectual faculties.

To sum up, this chapter was divided into four sections. The first was dedicated to a preliminary description of the objective sought by natural science. I outlined the basic desire to arrive at universal and necessary judgements. The second and third sections contained discussions on induction and falsification respectively. I defended the claim that a full understanding of both induction and falsification needs to refer to the essential link between, on the one hand, the purely logical discourse and, on the other hand, what people typically do in everyday situations. Having shown the inevitable significance of this link, I examined how the scientific method may affect the habits of the individual person. The two virtues I highlighted were reflective equilibrium and prudential risk-taking.

It is significant that the historical examples I used to illustrate the workings of the scientific method concerned grand universal theories, like Galilean Physics, Newton's Theory of Gravitation, Einstein's Theory of Relativity and so on. It is time now to broaden the horizon of my investigation. Natural science is not made up exclusively of grand universal theories. There are also local, everyday, scientific explanations that enjoy a high degree of plausibility in their own right. One may think, for instance, of how a ruler supported in the middle can be in balance with one coin on one side and two on the other. We explain this by saying that it happens because the one coin is twice as far away from the fulcrum as the two on the other side. Does the modesty of such an explanation show that it is not worth a place in a study of scientific method? If one tries to offer an account of science that caters exclusively for the grand universal theories and neglects the unpretentious local ones, one is bound to miss the everyday aspect of explanation.

The next step should therefore be to explore how scientific explanation is practised in everyday life. The hope is that this move may supply us with some valuable insights into the nature of explanation in general, and thus into the virtues associated with this more general intellectual activity.

Chapter 4

Explanation

Explaining something means making it clear, intelligible. Through explanation, we arrive at understanding. This is possible because some events are linked in a special way to other events. For example the event of kicking a football is linked to the event of it moving away. We explain the moving away by the kicking. Other events are explained in terms of conscious purposes, such as most human actions. We understand the kicking of the football by explaining why the person acted so. The kind of explanation characteristic of the natural sciences is different from everyday explanations as regards sophistication and detail. It is, however, quite similar as regards the overall desire to discover the links between the various events under consideration. In this chapter, I want to explore the way explanation affects our personal attributes. Not all personal attributes will be considered. My basic focus will be the effect explanation may have on those habits that involve doing good or evil. The very idea of such a project is already a daring enterprise because it goes against a well-established philosophical attitude according to which scientific explanation is completely detached from any morally significant attribute pertaining to the individual. The first step of my argument, therefore, will be to justify the possibility of a causal link between explanation and morally significant attributes. Once this possibility is secured, I will proceed by identifying some effects certain kinds of scientific explanation may have on the person's acting in line with genuine human flourishing.

The Main Objection

Some philosophers see scientific explanation as a purely intellectual enterprise. Indeed, the scene in recent scholarship has been set by the C.G. Hempel's influential proposal concerning what is to be explained, the explanandum, and what does the explaining, the explanans, conveniently situated within a valid syllogistic structure. The covering law is applied to a particular case so as to allow the explanandum to be drawn as a conclusion. The tendency here involves searching for necessary and sufficient conditions for a good explanation, determining the relevant elements of the logic of why-questions, and highlighting the paradigm argument for a good explanation. In all these operations, the hidden assumption is that any personal involvement of the one doing the explanation has no role whatsoever. Any relevance on practical matters explanation might have had in the past now shrivels away through a process of logical abstraction.

As this objection stands, however, it doesn't constitute a serious threat to my proposal. One may concede that there is a useful distinction to be drawn between the intellectual and the practical. This distinction has served philosophical reflection well in the course of history. One shouldn't, however, forget that thinking and acting are inseparable in everyday life. The operation we use to infer a logical conclusion from a set of premises is the same operation we use to infer how, when and why we act. Even Aristotle, who might be held responsible for separating intellectual from practical, was very much aware of the link between these two domains. He writes, for instance: 'But how is it that thought is sometimes followed by action, sometimes not; sometimes by movement, sometimes not? What happens seems parallel to the case of thinking and inferring about immovable objects. There the end is the truth seen (for, when one thinks the two propositions, one thinks and puts together the conclusion) but here the two propositions result in a conclusion which is an action – for example whenever one thinks that every man ought to walk, and that one is a man oneself, straightaway one walks' (*De Motu* VII, 701a8-12). The point therefore is that, although the Hempelian tradition has highlighted the intellectual side of the process of explaining, one should not forget that the process is a human one. It is carried out by agents who think and act, very often simultaneously.

This mild response may not be enough to convince a hard-line Hempelian. The objection, up to now, involved the simple claim that explanation might be considered an exclusively intellectual task. A more robust form of this objection can be envisaged not in terms of the task involved, but in terms of the results sought. When faced with an intriguing phenomenon, what we seek is its explanation, often assumed to be a unique discourse that may be partially or fully determined depending on our ability and technical facilities. Holding that, for any given explanandum, there is a full explanation waiting to be discerned by us is an avowal of strong realism of a somewhat Platonic kind. According to this new form of the objection, therefore, no causal links can be assumed to exist between explanation and personal habits simply because explanation is a process of tracing the contours of an intellectual landscape, as it were. There is no room for the idea that a correct explanation can be relative to the one formulating it.

This objection needs more careful analysis. Let us take a typical objector of this kind. David-Hillel Ruben defends the idea of a full explanation.[1] His distinction between full and partial explanation is quite simple to grasp. An explanation is partial when it omits certain relevant features; it is full when it includes all of them. It is clear, therefore, that for Ruben a full explanation exists even though we may not grasp it. The full explanation plays a similar role to a Platonic form, in the sense that the partial explanations participate in the full one in various degrees. With a logic that can be described as 'top-down', Ruben first sets his eyes, as it were, on a full explanation, assumed free of all significant relations to personal habits. He then proceeds by considering a sorting procedure by which some elements of the full

1 David-Hillel Ruben, *Explaining Explanation*, London: Routledge, 1992.

explanation are retained while others discarded. In this way, he wants to account for what people actually deal with in everyday life, namely partial explanations.

A more refined but similar working hypothesis is the one involving the so-called 'ideal-text', explicitly proposed by P. Railton in his refinement of Hempel's deductive-nomological model.[2] As in the case of Ruben, my interest is not on the validity or plausibility of the arguments presented, but on the assumptions underpinning the overall project. In this case, Railton's basic idea is to introduce the probability aspect within the normal DN model. In doing this, he makes use of the idea of an 'ideal text' of his deductive-nomological-probabilistic (DN-P) model of explanation. Fixing this ideal text as the objective allows him to employ the idea of a cumulative process approaching this ideal via partial contributions. He writes: 'where should we draw the line between explanation and non-explanation? The answer lies in not drawing lines, at least at this point, and in recognising instead a continuum of explanatoriness. The extreme ends of this continuum may be characterised as follows. At one end we find what I will call an *ideal DN-P text* [...], at the other end we find statements completely devoid of what I call "explanatory information".'[3]

The relevance of Ruben's and Railton's proposals should, I hope, be clear by now. Their way of engaging in an inquiry into the nature of explanation leaves no room for the causal link I am looking for. It leaves no room for any causal link between explanation and morally significant attributes, precisely because it leaves no room for any causal link between explanation and any personal factors whatsoever. Explanation is a matter purely of the intelligence. Any effects personal habits may have on explanation, or, the other way round, any effects explanation may have on personal habits, come out as irrelevant. The objective is to arrive at the one, ideal answer to the initial why-question. How could one ever conceive of having any significant links between explanation and virtue or vice?

Here lies, therefore, an apparently serious objection to my project. The response needs careful scrutiny of the word 'relevance' that is often used in this context, even by the authors mentioned above. It often remains in the margin of the argument and never thoroughly unpacked. Ruben, for instance, concedes that there is always a question of how we select from the full list of explanatorily relevant features so as to obtain the kind of explanation required. In this concession, he is suggesting that what we really use is not the unique full explanation, but an appropriate partial one. How we arrive at this partial explanation, he says, 'is a pragmatic audience-variant question'.[4] His predominantly top down approach, therefore, needs to be tempered with a bottom-up approach. By bottom-up I mean seeing explanation not primarily as an ideal text that would allow some approximations in different situations, but primarily as an operation in various particular contexts. In the first approach, we start with an idea. In the second, we start with a particular situation for explanation. In this latter case, some features are included and others discarded depending on

2 P. Railton, 'Probability, Explanation, Information', *Synthese*, 48 (1981): 233–56.

3 Ibid., p. 240.

4 Ruben, *Explaining Explanation*, p. 22.

what is already known, and depending on what is considered sufficiently satisfying. All explanations leave something out. But this leaving something out should not be taken to mean that there has to be a version of that particular explanation that does not leave anything out.

One may, to be sure, work out a fuller explanation from the partial one offered in the first instance. Ruben would probably agree that this may be done, for instance, by adding on the factors assumed known by the audience. In this sense, this approach is correctly called bottom-up. The crucial point, however, is that there is no guarantee that a fuller explanation, in the sense indicated here, is better than the original. It often happens, in fact, that the fuller an explanation becomes, the more ridiculous it gets. Think, for instance, of human action. The reason for acting is enough for an explanation. If I ask my sister why she's going to town, she replies that she wants to buy a pair of shoes. That's all there is to that kind of conversation. If she expands her answer by giving me a lecture on how neurones affect muscles, and how muscles constrict and extend, her reply would not be better but definitely worse – in spite of the fact that all she adds is true. Even if we avoid the explanation of human action, it is still true that sometimes a correct understanding of a given phenomenon involves giving up on certain levels of precision. Idealisation in physics is a case in point. For instance, we ignore air viscosity to arrive at the equation of the pendulum. There is epistemic loss if one limits oneself always to the finer theory. The loss arises because there are different levels of reality. The furniture of the world includes tables and chairs at one level, atoms and molecules at another. Truth is established by correspondence to different cognitive levels. For instance, the geometric shape of crystals is not found on the level of atomic physics but emerges holistically due to idealisations, such as 'surface', 'edges'. In a recent study on this point, Robert Batterman writes:

> It is not always true that the results – the principal features or 'laws' – arrived at by these methods are *merely* the first step toward a deeper understanding to be obtained when (and if) we make further strides in computation. There is a trade-off between 'improving' upon this (asymptotic) type of idealized result, by including more details in the computation, and achieving the understanding of patterns one desires. Such 'improvements' move one away from an account of the universality of the behaviour. The type of why-question shifts and the question about the existence of the pattern becomes less salient.[5]

---

5 Robert Batterman, *The Devil in the Details* (Oxford: Oxford University Press, 2002), p. 132. Another useful source for this discussion was: Fritz Rohrlich, 'Pluralistic ontology and theory reduction in the physical sciences', *British Journal for the Philosophy of Science*, 39 (1988): 295–312. Some are ready to accept ontological pluralism, e.g. tables, chairs on one level, atoms and molecules at another, but cannot accept the idea that all ontological levels have the same importance. For instance, Jaegwon Kim, in his paper 'Multiple realization and the metaphysics of reduction', *Philosophy and Phenomenological Research*, 52 (1992): 1–26, writes: 'Tables do not constitute a scientific kind; there are no laws about tables as such, and being a table is not a causal-explanatory kind. But that must be sharply distinguished from the false claim that there are no tables' (p. 25). This is misleading because the physical laws

It is obvious therefore that what Ruben calls a full explanation is often, if not always, unsatisfactory. The goodness of an explanation is not a matter of fullness but of fitness. It needs to fit well with the needs and aspirations of the audience as regards the amount and kind of information, and also given what is already known. An explanation can be defective by offering too much or too little of what is being sought.

One may perhaps want to defend Ruben by recalling what he himself included in his study various claims about the difference between a good explanation and a bad one. He mentions that an explanation could be bad because what was assumed known by the audience was not in fact known. It could be bad also if it conveys more information than is required. The right balance ensures that the explanation is a good one. He is therefore quite aware of the way explanations can be acceptable or unacceptable without reference to his assumed full, ideal explanation. So far, so good. My worries arise when he insists on distinguishing between the *goodness* and the *fullness* of an explanation. It is not difficult now, I hope, after the foregoing discussion, to see why this distinction is not tenable. There are no explanations in the abstract. Every explanation is situated within a context. And the major aim of the one offering an explanation is to offer a good one. No one thinks of offering a full explanation that may be bad in that context.

The thing to appreciate about this deficiency I see in Ruben's approach is that it brings to centre stage the variability within scientific explanation. Bas van Fraassen has highlighted this point by recalling how a given scientific explanation is always presented with respect to a given purpose. There are criteria that determine the correctness of the explanation, and other criteria that constitute the purpose. He writes:

> it is easy enough to say something true, impossible to say all that is true about a given subject. Selectivity in science is deliberate, purposeful and subject to evaluation as well. We ask not only if a given science provides accurate information about what is selected for attention but whether it has selected well, whether it answers all or many important or relevant questions.[6]

Given this multiplicity of criteria within the act of offering a scientific explanation, the conclusion is inevitable. It is wrong to consider explanation without referring to some attitudes or dispositions of the one offering the explanation, and also to some attitudes or dispositions of the audience receiving it.

It is because of this conclusion that my original project starts looking more promising. At the start, I stated the aim of this chapter as an exploration into the way explanation might affect our personal attributes of doing good or evil. What I have concluded at this stage is that accounts of explanation that refer to an ideal text or to

---

of free-fall, for instance, cover the behaviour of tables. The ontology of everyday common practice cannot be made redundant.

6 Bas van Fraassen, *The Empirical Stance* (New Haven: Yale University Press, 2002), p. 146.

a full explanation do not constitute a real threat. Further insight will be gained if we explore the domain of the pragmatics of explanation.

Pragmatic Aspects of Explanation

There is perhaps no better place to start an evaluation of how relevance criteria work in explanation than the work of Bas van Fraassen.[7] What I suggest is to analyse his arguments so as to have a clearer picture of how personal factors determine and are determined by explanation. First, a few words about why non-pragmatic accounts of explanation remain inadequate. Consider Hempel. He was clearly aware that one needs to include something about relevance even in his simple DN model. For there to be an explanation, he claims, we must have both explanatory relevance and testability. By the former, he means that the explanatory information adduced affords good grounds for believing that the phenomenon did, or does, indeed occur. By the latter, he means that the statements constituting a scientific explanation must be capable of empirical test. In spite of the logical austerity these two points seem to manifest, they do not constitute definite necessary and sufficient conditions for explanation. For instance, against the first point about explanatory relevance, one can think of examples where giving good reasons for believing that *p* does not always amount to answering the question: Why *p*? To take Aristotle's example in the *Posterior Analytics*, I believe the planets are near because they do not twinkle. They are not near *because* they do not twinkle. The twinkling does not explain why they are near. Hempel's account, therefore, whatever its merits, cannot be said to need no revision.

Indeed, revision is needed because of the two major difficulties any quick account of explanation is obliged to face some time or other. The first has to do with asymmetries. Consider, for instance, the consequences of a particular scientific inquiry. This very often takes the form of a statement that one condition obtains when, and only when, another does, a little like Aristotle's observation that, for heavenly bodies, twinkling and proximity go together. Another very simple case: when the sun is at an elevation *E*, the tree's shadow is of length *L*. Such a concluding statement is ambiguous as regards explanation because one condition explains the other but not *vice versa*. In this case, such asymmetry occurs because the elevation of the sun

7 Especially Bas van Fraassen, *The Scientific Image* (Oxford: Oxford University Press, 1980), chapter 5. An important forerunner, who deserves a mention, is Michael Scriven, who at an early stage challenged the standard distinction between explanation and description. He argued that explanation is not a matter of something other than description but a matter of the context in which that description is placed. The DN model's insistence on truth should be augmented with the insight that the covering laws in the explanans must have a role in systematising our knowledge. See: M. Scriven, 'Explanation, Prediction and Laws' in H. Feigl and G. Maxwell (eds), *Scientific Explanation, Space and Time* (Minneapolis: University of Minnesota Press, 1962), pp. 170–230; M. Bunzl, *The Context of Explanation* (Dordrecht, London: Kluwer Academic Publishers, 1993).

explains the shadow's length. But, even though we can deduce *E* from *L*, we cannot say that the shadow's length explains why the sun is at that particular elevation. The second difficulty has to do with explanatory closure. Simple Hempelian-type accounts of explanation do not allow any stopping point where why-questions cannot be asked any longer. One seems free to ask a why-question about any initial singular case; then another why-question about the covering regularity that explained that singular case; then another why-question about the covering law that explained that regularity that explained that singular case, and so on. Explanation has no boundaries. This, however, is not what we see in actual practice. There are spectacular cases where explanation is rejected in the name of genuine progress. Galileo simply rejected the Aristotelian's why-question: why does a body free of impressed forces retain its velocity? Likewise nowadays, we reject the why-question: why does this particular uranium atom disintegrate just at time t_0 rather than at a later time t_1? This is a well-formed why-question similar to others which are acknowledged as answerable. It is however rejected because the theory within which it is situated is considered complete.

What then should a more responsible account of explanation look like? One needs to start at the beginning. What we are doing in the act of explaining, at least in the natural sciences, is determining a causal chain that leads up to the event. What we start with are events that are enmeshed in a net of causal relations. Imagine these relations as lines joining events, criss-crossing in very complicated ways. Explanation of why an event occurs consists typically in determining the notable factors in the part of the causal net formed by lines that lead up to that event. Given the complexity of the causal net, we cannot assume that there is any one-one correspondence between events and causal relations. A single event may be explained in various ways. Norman Russell Hanson offers a clear example:

> Consider how the cause of death might have been set out by a physician as 'multiple haemorrhage', by the barrister as 'negligence on the part of the driver', by a carriage-builder as a defect in the brake-block construction', by a civic planner as 'the presence of tall shrubbery at the turning'.[8]

Notice how the discussion on explanation and cause has suddenly been brought down from the ethereal heights of abstract thought to the level of everyday living, where contexts differ according to various needs. The cause that constitutes the explanation is indeed one particular line of the causal net. Acknowledging that particular line as an explanation is equivalent to acknowledging one's view-point. Van Fraassen explains this fact by saying that the sentence '*A* is the cause of *B*' has a meaning within a practice where the term 'cause' has a specific use. Both the nature of *A* and the nature of *B* are important. The major importance, however, is not the nature of *A* and *B*, but the orientation or the chosen point of view of the speaker that determines what the word 'cause' is used to signify. This is not a recent discovery. Aristotle's

8 N.R. Hanson, *Patterns of Discovery* (Cambridge: Cambridge University Press, 1958), p. 54; quoted also in van Fraassen, *The Scientific Image*, p. 125.

four causes are indeed an indication of the various interests that can determine the selection of a range of relevant factors for a why-question. Why do our hearts beat? One explanation could be in terms of efficient causes, involving, say, the flexing of muscles. Another could be in terms of final causes, involving the pumping of the blood, the supplying of oxygen and the survival of the entire organism.

It is, I hope, clear by now why van Fraassen's diagnosis of the limitations of context-independent accounts of explanation is important for my overall project. The discovery that the personal view-point of the one offering the explanation makes a difference is crucial. It allows further exploration in the area corresponding to the theme of this chapter, namely the area of the possible interaction between scientific explanation and personal morally-relevant attributes. As I mentioned above, the important term here is 'relevance'. The explanation offered must be relevant to the audience to whom it is directed. Some further analysis of this term 'relevance' will facilitate my task of determining what aspects of explanation have a bearing on the life of virtue.

Van Fraassen explores relevance in terms of the relation between the initial why-question and a proposition, a candidate for a reply. This proposition would qualify as an explanation, as a correct reply, if it is in a certain relation of relevance with the why-question. In a more precise way, he starts by investigating the nature of the initial why-question Q. This question is determined by three factors:

the topic P_k,
the contrast-class $X = \{P_1, P_2, ..., P_k, ...\}$,
the relevance relation R.

The first factor is simply the subject matter of the question. The contrast-class is a set of propositions that sharpen the main subject of the question by indicating more precisely the nature of the explanation required. For instance, consider the question: why does smoking increase the probability of developing lung-cancer? This is of the form: why P_k? We specify further by adding a contrast: why P_k rather than P_1? In this example, P_1 could be: *increase the probability of developing another lung ailment.* The full explanandum is thus clearer: why does smoking increase the probability of developing lung-cancer rather than increase the probability of developing another lung ailment? This more specific description is not the same as the original with *another* contrast. Consider P_2 as: *not increasing the probability of lung-cancer*. The original why-question with this contrast is not the same as before. It becomes: why does smoking increase the probability of developing lung-cancer rather than not increase it at all? It is clear, therefore, that the contrast-class has a very important role in explanation. The third factor mentioned by van Fraassen is the relevance relation. This factor corresponds directly to what is already known by the audience. If a given why-question is further specified by contrast, it can still be answered in various ways depending on two factors. It depends on what the explainer takes for granted as obvious, and what the explainer thinks his audience already knows. For instance, for the above question about smoking, a correct explanation depends on whether the questioner is a medic or a school-boy.

Hence, the why-question can be considered in general as a triple

$$Q = < P_k, X, R >.$$

Van Fraassen's original contribution here is not so much this general form of the why-question. It lies rather in the introduction of the role of background theory and background information. The correct explanation sought is determined not only by the main content of the why-question together with its contrast-class, but also by the context in which it needs to fit well, as it were. One cannot neglect, therefore, the body of accepted knowledge and information that depend on the explainer, and the audience. There may even be particular why-questions that lose their meaning if transported from one context to another. This point corresponds to the condition of explanatory closure mentioned above. Galileo's way of doing natural philosophy was so different from the Aristotelian way that Galileo's background theories simply excluded the why-question: why does a body free of impressed forces retain its velocity? Van Fraassen considers this aspect of the context-dependence of explanation as the most important aspect for a correct understanding of this theme. He writes:

> the discussion of explanation went wrong at the very beginning when explanation was conceived of as a relationship like description: a relation between theory and fact. Really it is a three-term relation, between theory, fact, and context.[9]

The main point here is that this approach appreciates what may be called the ontological richness of the events to be explained. A given event may be seen in various ways, all of which are true depending on the 'explanatory position' of the one doing the explaining. This idea is essentially an extrapolation from visual experience, where one and the same landscape can be perceived from different perspectives corresponding to each observer. What I'm calling 'explanatory position' refers to a point in context space. Just as in the visual case we assume the existence of a space consisting of the set of all possible positions the observer may take, so also for van Fraassen's account of explanation. Here we are invited to think of a context space. This is the set of all possible contexts the explainer and the audience can be situated in.

So far, our line of argument seems quite straightforward. There is, however, an element of weakness. The introduction of the context apparently means that we are sliding uncontrollably into relativism. The relevance relation seems like a door opening up to the attitude of anything goes. Consider an example involving a set of background beliefs that are nowadays unacceptable, such as the set of beliefs associated with astrology. Why did J.F. Kennedy die on 22 November, 1963?[10] A possible answer is 'because the sun, planets, moon and stars were in such and such position'. This explanation consists in a *true* description of the position of astronomical bodies at the time of Kennedy's death. The explanation however is

9 van Fraassen, *The Scientific Image*, p. 156.

10 The example is from W. Salmon, 'Four decades of Scientific Explanation' in P. Kitcher and W. Salmon (eds), *Scientific Explanation* (Minneapolis: University of Minnesota Press, 1989), pp. 3–196.

unacceptable. The reason is that, in the context of twentieth-century science, there is no such thing as astral influence. Descriptions of celestial bodies are not relevant for the understanding of most events on Earth. Van Fraassen seem to neglect such examples. He doesn't seem to realise that, if we accept just any relation as relevant, there would be no limits to what could count as a correct explanation. He needs to specify which kinds of relevance relation is to be included in the triad

$$Q = < P_k, X, R >.$$

This criticism of van Fraassen's position has been expressed perhaps most clearly by P. Kitcher and W. Salmon.[11] They see his theory of explanation as part of an effort to show that theoretical virtues beyond the saving of the phenomena are pragmatic. This project, they insist, gives rise to the dangers of what they call the 'anything-goes' theory of explanation. What van Fraassen lacks is objective relevance criteria that can overcome the traditional problems of the theory of explanation. Otherwise his proposal of the explanation triad remains, as it were, open at one end. They therefore conclude:

> if van Fraassen avoids the Scylla of the 'anything goes' theory of explanation, then he is plunged into what he would view as the Charybdis of supposing that there is an objective virtue of theories distinct from their salvation of the phenomena. From our perspective, Scylla is (to say the least) uninviting, but Charybdis feels like the beginning of the way home.[12]

The Kitcher-Salmon objection is easy to block. Sympathisers of van Fraassen's pragmatic approach can recall how the astrology example is much too simple to be generalised. History teaches us to be wary of believing in clear and abrupt changes in relevance-criteria. Defenders of van Fraassen can concede that astrology-related relevance-criteria are indeed quite distinct from our present ones. But, even for this example, humanity's struggle to move from one set of criteria to the other has been a long one. There were long periods of mixed criteria. Other examples are certainly not any simpler. The point is that the Kitcher-Salmon objection underestimates the difficulty any group of people must face in the act of distancing itself from its own background beliefs. The pragmatic approach need not resort to subjective criteria of relevance. It is not an account of explanation that boils down to the idea that an explanation is good if it is good for me. What determines the relevance relation is a set of *objective* criteria associated with the particular context of discourse being engaged in. Using the Latin *qua* can help to grasp this point. I can explain, say, the burning of a candle *qua* chemical change, and I can explain the same phenomenon *qua* liturgically significant symbol. In each of these explanations there are criteria

11 P. Kitcher and W. Salmon, 'Van Fraassen on Explanation', *Journal of Philosophy* 84 (1987): 315–30.

12 Ibid., p. 330. For further developments of these arguments, see Elisabeth A. Lloyd and Carl G. Anderson, 'Empiricism, Objectivity, and Explanation', *Midwest Studies in Philosophy*, 18 (1993): 121–31.

of relevance that are objective, in the sense that they go beyond the purely personal arbitrariness.

Even within the strict limits of the natural sciences, the various possible perspectives on the same phenomenon are indeed a manifestation of the various relevance-relations that can be associated with a single explanandum. And these criteria are objective. Consider, for instance, the melting of the polar ice-caps. Meteorologists study this phenomenon from a physical and climatological point of view, whereas zoologists study the same melting of the ice-caps from a zoological point of view, and ecologists from the ecological point of view. The data is selected according to their various points of view. The data that is relevant for a physicist is different from the data that is relevant to the interests or concerns of a zoologist or of a climatologist. Different relevance-relations are at work here. None are exclusively subjective. The explanations differ from context to context for two reasons. One the one hand, there are factors related to the background knowledge of both questioner and audience. On the other hand, there are also factors related to the ontological richness of the explanandum, which allows it to be appreciated in more than one way. It may be useful to recall here that, for those who are tempted to hold that ontology is a seamless whole that forbids distinctions between levels, it is enough to recall that macro-regularities in nature can be realised by wildly heterogeneous lower level mechanisms. From the point of view of the lower level theory, many of the microstructurally relevant details are irrelevant for the upper level phenomenology. What I'm calling ontological richness, therefore, is multi-layered. It necessitates different taxonomies that are all valid, each with its associated science.[13]

The significance of these conclusions for my overall project should now be recalled. This last section has been dedicated to the pragmatic element within explanation. It is clear by now, I hope, that context-dependence in the area of scientific explanation is inevitable. It is the context that determines the nature of the relevance relation included in van Fraassen's triad $Q = < P_k, X, R >$. There is a major objection to this way of understanding explanation. It consists in saying that bringing in the context would undermine explanation. It would open the door to subjectivism and relativism. To reply to this objection I recalled that relevance need not be associated with the whims of the individual. On the contrary, relevance has an objective aspect because, on the one hand, it is related to the epistemological state the community of inquires is in, as regards its background knowledge and its basic, axiomatic beliefs; on the other hand, it is also related to the ontological richness of the explanandum that allows different levels of inquiry. So context-dependence is inevitable. This fact undermines the idea that explanation can be seen as a discourse lying beyond all perspectives. This erroneous idea has been instrumental in destroying all interest in the question regarding the possible interaction between the way people explain

13 More detail on this point can be found in: J. Fodor, 'Special sciences, or the disunity of sciences as a working hypothesis', *Synthese*, 28 (1974): 97–115; R.W. Batterman, 'Multiple realizability and universality', *British Journal for the Philosophy of Science*, 51 (2000): 115–45.

things and the way they live. Now that context-dependence is seen as inevitable, we can safely proceed to explore the way the scientific mentality may have an impact on moral character.

Explanation and Virtue

Any effect on the moral character is discernible only in the long run. The individual explains one phenomenon one way, explains another phenomenon the same way, and then a third the same way, and so on. A habit sets in. At the surface, the habit deals with ways of explaining things, but, given the unity of the person, we cannot assume that such an explanation-related habit is totally detached from other habits. In fact, there is a specific word used to refer to such habit-formation, especially when considered not merely in relation to the individual, but in relation to a group of individuals, a society. The word is mentality. I am taking the term 'scientific mentality' to mean, in broad strokes, the set of habits that distinguishes a particular society before the predominance of scientific explanation in that society's intellectual efforts from the same society after it. Since it is easier to consider the individual, however, I will not deal with society at large but will concentrate mainly on the individual. I will proceed, as in previous chapters, with an inquiry into the two major negative effects scientific explanation may have on personal dispositions. One such effect is the disposition to neglect the importance of context-dependence of all explanation. The other effect is the disposition to exaggerate the importance of such context-dependence. I will consider each of these cases in turn. These two dispositions will be shown to lead to opposing vices. Each represents an extreme that does not square with genuine human flourishing. The virtue associated with the area of explanation will then become determinable as the mean between these two extremes.

First, then, the disposition to neglect the importance of context-dependence of all explanation: this arises from attitudes related to the Hempelian approach as described above. The crucial point is the misleading conviction that there is one full explanation for a given explanandum. Such a conviction sets in when the relevance relation is neglected. For instance, a given phenomenon is considered fully explained by the explanans in the form of the DN scheme. The person who adopts this way of explaining may eventually become more and more used to it to such an extent that other possible perspectives on the issue are forgotten or deemed impossible. This essentially means that the person becomes blind to the other kinds of legitimate explanatory discourse determined by other relevance relations. What is happening here can be understood in terms of priorities. One starts with the situation where various perspectives are considered possible as regards a particular explanandum. One starts with a particular situation, for instance the melting of the polar ice-caps, referred to already in the illustrative example mentioned above. As regards this melting of the polar ice-caps, some adopt a physical point of view, others a zoological point of view, others an ecological point of view. Different relevance-relations are at work here, and none of these relations are deemed better than the others, in any absolute sense. This is all well and good. The problem starts to surface when one of the relevance relations is given priority. What happens most commonly is that the physics point of view carries the day. The explanation related to physics is considered the best, the fullest, or simply the only correct one.

In this line, therefore, the real danger is reductionism. What we are practically saying is that, as explanatory context-dependence is let to slide towards a minimum, the tendency of reductionism increases. Reductionism is often considered a position definable within the limited realm of natural science, as when biology is held to be reducible to physics. Sometimes, reductionism is even discussed within the more limited realm of physics only, as when we say that the gas law $PV = nRT$ can be reduced to the kinetic theory. In the context of the present discussion, however, we are in fact considering a kind of reductionism operative not only between theories but between disciplines in a more general sense. Hence, the tendency to engage in reductionistic explanation includes the tendency, or set of attitudes, referred to by the more ideologically sounding term 'scientism'.

Scientism comes in various kinds. Three of the more important kinds are the following.[14] Methodological scientism involves the extension of the use of the methods of natural science to other disciplines. This extension is done in such a way that these methods exclude previously used methods considered central to these disciplines. Epistemic scientism is another kind. According to this, the only reality we can know anything about is the one natural science has access to. Another dimension of scientism has to do with values. What is sometimes called axiological scientism is the position according to which science is the only truly valuable realm of human life; all other realms are of negligible value. These three dimensions of scientism give clear indications that, as an ideology, it involves deep habits of thinking and living. These habits all express one fundamental position: that science alone can and will eventually solve all the genuine problems of humanity. Some philosophers have been pioneers in investigating where this position may lead. Auguste Comte, for instance, has become famous for his idea that the age of natural science has finally dawned. As a consequence, he claimed that all other kinds of discourse, such as religious or metaphysical discourse, need to be considered redundant. The inherent instability of such an attitude became obvious through the work of the Vienna Circle. The very statement describing the exclusivity of scientific explanation can never itself be part of a scientific explanation. The entire scientistic enterprise is founded on a claim that the enterprise itself is obliged to consider irrelevant. The major inconsistency, therefore, lies in the fact that scientism, sooner or later, involves sawing off the very branch on which one is sitting.

These familiar arguments confront scientism as a theory. I will argue at a different level. I will argue that scientism is unacceptable not because it is self-contradictory when considered in all its detail, nor because it is inconsistent with other very basic beliefs we cannot easily sacrifice. I will argue that scientism is a doctrine embedded

14 I am indebted here to T. Sorell, *Scientism* (London: Routledge, 1991); M. Stenmark, *Scientism: Science, Ethics and Religion* (Aldershot: Ashgate, 2001).

in a way of life, a theory within a practice. Hence, scientism can be evaluated at the level of action. The idea here is to show that scientism is to be avoided because, taken in its broader sense involving practice, it is a vice. In other words, it is to be avoided because there are good reasons to believe that it engenders an attitude that, in the long run, hinders the genuine flourishing characteristic of human beings.

Now for the reasons: the argument can be conveniently presented in three steps. Step one involves expressing clearly the moral consequences of scientism. A life marked by scientism, which, as I mentioned above, includes reductionism, involves making decisions about action with exclusive reference to natural science. This means, in effect, that ethical norms and beliefs are reduced to — or taken to be reducible to — principles that are purely scientific. The pride of place in these considerations goes not to physics but to biological evolution. A scientistic life-style is one in which the individual proceeds from a knowledge of the material, evolutionary basis of moral feeling to generally accepted rules of conduct. Of course, a clear distinction between *is* and *ought* is denied. Any description of the morally good is taken to express behaviour patterns that have been naturally selected in the course of the long millennia of evolution because they favour survival of the species. Any description of acts that are morally evil is taken to express behaviour patterns that have been naturally selected because they endanger survival of the species. A clear example of this is offered by M. Ruse and E.O. Wilson:

> One of the most instructive cases is provided by the phobias. These intense reactions are most readily acquired against snakes, spiders, high places, running water, tight enclosures, and other ancient perils of mankind for which epigenetic rules can be expected to evolve through natural selection. In contrast, phobias very rarely appear in response to automobiles, guns, electronic sockets, and other truly dangerous objects in modern life, for which the human species has not yet had the time to adapt through genetic change.[15]

The major feature here is that no distinctly moral qualities are assumed to exist.

From here, I move on to step two of my argument. Endorsing scientism means living in a world where moral norms are all understandable on the model of phobias. In other words, a typical maxim such as 'Thou shalt not kill', as an expression, has nothing more behind it, or inherent within it, than the similar expression 'Beware of snakes'. The mechanism behind the emergence of such expressions is the same. What we call moral sanctions, like the former, are nothing more than sophistically expressed phobias. According to this view, therefore, morality is considered an illusion. Moreover, since the very idea of illusion gains credibility through the identification not only of the deceived but also of the deceiver, a fashionable way of describing this particular aspect of current sociobiology is to identify the deceiver, surprisingly enough, as our own genes. The proposal is that we, as individual persons, are being deceived continuously by our genes. In the words of Ruse and Wilson:

15 M. Ruse and E.O. Wilson, 'Moral philosophy as applied science', *Philosophy* 61 (1986): 183.

> We do not claim that people are either unthinking genetic robots or that they co-operate only when the expected genetic returns can be calculated in advance. Rather, human beings function better if they are deceived by their genes into thinking that there is a disinterested objective morality binding upon them, which all should obey.[16]

The next step in my argument is to uncover the serious problems with this kind of reasoning. The entire argument is, in fact, objectionable on various levels. There is, to start with, a grave logical problem with the idea of genes being capable of deceiving others. Attributing intentions to genes involves a conceptual mistake. It is not difficult to see that we have here an extreme form of the mereological fallacy mentioned already in Chapter 2. In that chapter, the problem was the one of ascribing a psychological predicate to a part of the human person, like the brain, when in fact the correct use of such predicates requires that they be ascribed only to the person as one individual. Put bluntly, the one who thinks is the person, not the brain. In the genetic reductionism we are dealing with in this chapter, the same kind of confusion in the use of psychological predicates occurs. To deceive someone means to have the intention of depriving him or her from the whole truth concerning a particular matter. This can only be carried out by individuals with minds that intend. Genes are parts of persons. They can only be said to think, intend, conspire and deceive in a very loose metaphorical sense. And such loose senses are clearly inappropriate for clear philosophical argument. One cannot just shift any psychological predicate from where it really belongs to an imaginary 'small person' presumably situated within each and every gene of the real person.

This is the logical problem. Ruse and Wilson's idea is also fatally flawed on another front. It is fatally flawed as regards the practical aspect of the scientism that their idea represents. To see this, consider first how numerous are the kinds of duties that are woven into the very fabric of any individual's everyday living. Indeed, duties are a real aspect of any individual's life not only in the occasional situation where the individual faces a deep moral dilemma. They are a real aspect in nearly all situations. They are expressed in various principles that are then translated, as it were, into the ever-changing particularity of everyday living. 'Thou shalt not kill' implies refraining from exacting vengeance by doing harm to others at the office, in the home, in this particular squabble, after that particular hurt. If scientism results in a life-style that takes morality to be a vast deception, the problem runs very deep. We are not dealing with occasional moments of deception, but with an entire immersion into inescapable falsehood determining most of what we do in everyday life. Scientism, therefore, as expressed by Ruse and Wilson's idea, places the individual in a world of lies. It is reasonable to draw the conclusion from here that the state scientism ends up in is one where no genuine happiness is conceivable. It does not make sense to say that people who know that they are being continuously deceived can also flourish in a genuine way. As we may recall, however, scientism consists essentially in a kind of explanation that is devoid of any idea of context dependence. It consists in a kind of

16 Ibid. p. 179.

explanation where only one correct viewpoint on things or events is assumed to exist and where only physicalism matters. The case of scientism, therefore, shows how neglecting context-dependence is bad for the individual. It is associated with a life situation that blocks genuine human flourishing.

The point and relevance of this conclusion for my overall argument should now be recalled. The major starting point for this section was the conclusion drawn from the previous section, namely that the context-dependence of scientific explanation is inevitable. Here the argument can branch out in either of two directions. On the one hand, we may explore the consequences of neglecting context-dependence. On the other hand, we may also explore the consequences of exaggerating such dependence. The investigation up to now dealt with the first branch. I have essentially shown that neglecting context-dependence does not square with genuine human flourishing. Neglecting context-dependence amounts to reductionism. And reductionism, considered not superficially merely as a thesis but more substantially as a practice, is equivalent to scientism. From here, I advanced the argument that scientism is incompatible with genuine human flourishing. The main reason was that, as regards the ethical sphere, it implies living in a world of illusion. The overall conclusion, therefore, is that whenever the scientific mentality reinforces an attitude of context-dependence in explanation, scientism is involved. Scientism makes itself felt as a way of life, as a basic habit, or set of habits, that determines much of the general behaviour of many of those engaged in natural science. The arguments I advanced show that such a set of habits has all the characteristics of a vice.

This therefore is the vice of neglecting the context-dependence of scientific explanation. It constitutes one direction I intended to explore; now for the opposite direction. Instead of letting context-dependence shrink to zero, we can consider the case when it grows to infinity. In such a situation, we have an exaggerated importance given to context-dependence. The argument I want to present is that such an exaggerated context-dependence leads to vice because it impedes the individual's acting in line with genuine human flourishing.

The best way to start is to consider what it means to exaggerate the context-dependence of explanation. A person prone to this kind of exaggeration will typically tend to be the opposite of an adherent of scientism. Scientific explanation becomes just one of the infinitely many kinds of discourse that can be legitimately engaged in as regards a particular state of affairs. Whenever such a person encounters a scientific explanation *E*, even when this explanation is supplied by the person himself or herself, what is underlined is the fact that *E* is valid for a very specific context. Exaggeration of context-dependence means that the validity, or appropriateness, or relevance of *E* is taken to be extremely sensitive to context-change. If the context changes, even slightly, then *E* will not remain valid. The major feature here is this extreme sensitivity to context-change. Consider an example. Take the shrinking of the polar ice-cap as the phenomenon under investigation. A typical explanation could be: 'the shrinking of the polar ice-cap is due to the emission into the atmosphere of great quantities of carbon dioxide'. For the person who exaggerates context-dependence, the feature of this explanation that is highlighted most is not its intrinsic plausibility but its dependence on a particular view-point. Such a person would say: this explanation is relevant for the climatologist's interests. As regards the zoologist, say, a valid explanation would involve a totally different set of parameters. It would involve patterns of increase or decrease in the life of flora and fauna. The two explanations are indeed compatible and mutually supporting. The person who exaggerates context-dependence, however, will always highlight their difference and mutual autonomy rather than their compatibility.

As can ready be seen, what we have in this example up to now is just two explanations of the same phenomenon. They are distinct from each other not because of the subject matter, but via what van Fraassen calls the contrast class and the relevance relations. Now consider the case when, for a given subject matter, the number of variations due to changes in the relevance relations increases steadily from two, as in the above example, to a considerable amount. The result will be that the alleged superior value or significance of any one individual explanation diminishes. The person who exaggerates context-dependence reasons this way: 'The account involving the emission of carbon dioxide is fine, but there are so many view-points one can adopt – each with its own distinct, autonomous and self-contained explanation – that the desirability of that particular account diminishes considerably. The account is true, yes, but, if one sees the subject matter from another view-point, things look differently.'

The consequences of adopting such an attitude are not difficult to recognise. Scientific explanation starts looking very much like a never-ending task of refinement over refinement, with the inquiry bifurcating endlessly into ever more specialised contexts of explanation. Any concern for an integrated understanding of the whole evaporates slowly. In the long run, what results is a neglect of what may have an impact on the life of the person as a whole. The individual's intellectual output is splintered, as it were, into various contexts, and no space is left for a principled reflection on how the person's life, taken as a whole, can be directed towards the good. Even the very question starts looking like something either too difficult to handle or even utterly meaningless.

It might be objected here that this character trait associated with exaggerating context-dependence, taken as I have described it, does not really amount to a vice. It apparently does not involve acting in ways that go against genuine human flourishing. What we have here is merely an element of neglect. The person becomes so much engrossed in the context-dependence of any one explanation that he will lose touch with any general attributes of his life as a whole. There is certainly something to this objection, but there are various possible replies. The simplest is perhaps to concede that the character-trait I have described can only be called a vice in a general sense. It merits this designation because it hinders the person from confronting questions that are essential. It makes the person, as it were, miss the wood for the trees. Admittedly, it would have been a more serious kind of vice had it involved attitudes more specifically directed against genuine human flourishing, rather than just neglect. But the fact that it could have been worse as a vice does not imply that it isn't a vice of a less direct kind.

There are sufficient reasons, therefore, to consider our original task done. What I had fixed as my aim was the identification of two opposing vices associated with explanation. Now it is clear that the first one involves minimising the importance of context-dependence of scientific explanation; the second involves maximising the importance of such context-dependence. This pair of opposing vices allows me now to apply, in a simple and quick way, the Aristotelian method of seeking the mean. As in previous chapters, the idea is to determine, to some extent, a particular virtue associated with explanation. It is not difficult to see that, as regards the topic explored in this chapter, the good habit enhanced by the right kind of scientific explanation corresponds to an attitude that is balanced in the sense that it gives the right amount of importance to the context or view-point. In other words, this amounts to saying that virtue lies in showing the right kind of respect towards scientific explanation, in so far as it is *located*. The virtuous person does this without losing sight of other kinds of possible, principled inquiries, especially the kind of inquiry that deals with the life of the individual taken as a whole.

To summarise this chapter, let us now recall the overall line of argument. I started with blocking an objection. Scientific explanation is often assumed to be an abstract matter unrelated to the personal attributes of the one engaged in it. I refuted this by showing that there is indeed the possibility of causal links between explanation and morally significant attributes. Accounts of explanation that are paraded as abstract in this sense employ the idea either of an ideal text or of a full explanation. I showed that neither of these constitute a real threat. This was the main thrust of section one. Section two was about pragmatic features of explanation. I showed essentially that context-dependence in the area of scientific explanation is inevitable. It is the context that determines the nature of the relevance relation included in van Fraassen's triad $Q = < P_k, X, R >$. I observed that this may not convince everyone. An objector may insist that bringing in the context would trivialise explanation. It would open the door to subjectivism and relativism. To reply to this, I recalled that relevance need not be associated with the whims of the individual. On the contrary, relevance has an objective aspect. It is related to the background knowledge of the community of inquires where the explanation occurs. It is also related to the ontological richness of the explanandum that allows different levels of inquiry. So context-dependence is inevitable. The question of how explanation might affect the way the individual lives, therefore, becomes respectable. From here, I proceeded by identifying how genuine human flourishing is hindered either by neglecting the context-dependence of explanation or by exaggerating it. The virtuous person is the one who gives the context-dependence of scientific explanation with the right amount of importance.

By now, it is becoming clearer, I hope, that a rich understanding of natural science must involve the interconnectedness between its methods, its ways of explaining, the background beliefs engraved within the cultural context within which it flourishes, and the habits of the individuals who engage in it. We cannot expect such an understanding to allow formulations with mathematical precision. The immense number of variables involved precludes this. We should not, however, draw the conclusion that a reflection of this kind is thereby rendered futile. To each kind of inquiry, its own degree of precision. With this proviso in mind, it is time now to take a further step. What happens to the entire fabric, often referred to as the scientific mentality, when there are significant changes, such as when a major theory is replaced by another? An examination of such changes will shed new light on how virtues and vices of individual scientists can be affected by their discipline.

Chapter 5

Science and History

As scientific inquiry proceeds through time, crucial periods occur when the correct theory must be chosen from among various contenders. A theory, in the most general sense, can be taken to be a system of ideas bringing various strands of our knowledge together or explaining some set of related events. When several possible systems of ideas are proposed to explain the same set of events, the scientific community seeks clear criteria of choice. Is it always possible to find such criteria of choice? An affirmative answer to this question, in line with a certain amount of optimism as regards human rational abilities, holds the promise of arriving at a complete list of methodological principles, presumably containing all there is to know about inductive logic and probability. A certain security is therefore associated with this affirmative answer. An answer in the negative would severely undermine such security and would affect some very profound attitudes of anyone marked by the scientific mentality.

The last few decades have witnessed a remarkable increase in interest in this issue. A certain revolutionary aspect of scientific development has been discussed at length. What I propose is to examine, in a first section, the main consequences of such revolutionary aspects. I will be defending the thesis that a fully rational list of criteria to choose the correct way how to proceed through major upheavals in scientific thinking is a fiction. Stated in another way, the way forward in such upheavals will inevitably involve personal value judgements that may originate from politically or socially contingent factors. There is a certain kind of practical wisdom involved. In the second section of this chapter I will explore how this dependence is in fact not a one-way dependence but a two-way dependence. On the one hand, personal attitudes and dispositions affect some crucial periods of the history of science. On the other hand, the historical awareness introduced in science studies affects personal attributes. It is here, in this latter kind of dependence, that the historical consciousness and the life of virtue may be explored. Given that the search for the best explanation forms an integral part of the scientific mentality, I am assuming that prolonged exposure to this kind of practice, especially with the heightened awareness of history that it now involves, will enhance certain dispositions of the individual. An analysis of this aspect will enable me to determine which virtues and which vices are associated with this particular perspective on natural science.

Revolution and Personal Value Judgements

The claim I want to defend in this section is that there are genuine revolutions in science and that they inevitably involve personal value judgements. My reasoning will be presented in four steps. The first step is to explore the most radical account of scientific revolutions. I start here because according to this radical account, as we shall see, personal value judgements in scientific revolutions are not merely present; they constitute the *only* determining factor in change. The following steps of my argument, then, will deal with an objection and with a less radical account of scientific revolutions. It will become evident, I hope, that in whatever way we account for scientific revolutions, there are always personal value judgements involved.

I start then with the most radical account. This is the product of a shift in interpretative priorities. From an emphasis on logical analysis, including detailed evaluations of induction and falsification, we move to an emphasis on historical studies. From how matters should be we move to how matters are. The shift in priorities requires us to think in terms of large, complex units and avoid over-simplifications. The very notion of a theory is too narrow. Theories are linked to each other in a very complex ways. When seeking to evaluate what happened in a definite historical period, or what is happening in our own time, we cannot easily isolate a given theory from its context. From theories as units of understanding, therefore, we move to paradigms: 'universally recognised scientific achievements that for a time provide model problems and solutions to a community of practitioners'.[1] The term 'paradigm' is used to refer not only to one specific scientific theory and to other theories closely related to it, but also to the very way of selecting and treating the observations that support the theory, together with the criteria for accepting a scientific explanation as relevant and acceptable. The paradigm guides normal science as it solves various puzzles that extend its explanatory range. Normal science corresponds to the kind of work done by those who never question the basic theories that characterise their paradigm. In face of contrary evidence, they consider that their core theory beyond question. Any contrary evidence will eventually be explained when finer adjustments are made to it. Consider for instance Newtonianism and

1 T.S. Kuhn, *The Structure of Scientific Revolutions*, 2nd ed. enl. (Chicago: University of Chicago Press, 1970). Many associate the term 'paradigm' exclusively with Kuhn. It had however a long and significant history before the publication of this book. Some historians have identified an important link between L. Wittgenstein, N.R. Hanson and T.S. Kuhn. Hanson concentrated on the basic experience of perception and started with Wittgenstein's point that: 'seeing is not only the having of a visual experience; it is also the way in which the visual experience is had.' (Hanson, *Patterns of Discovery*, p. 15; Wittgenstein, *Philosophical Investigations*, p. 196). Kuhn's term 'paradigm' is taken from Wittgenstein (*Philosophical Investigations* §§ 20; 50 55; 57). See also M.J.S. Hodge and G.N. Cantor, 'The Development of Philosophy of Science since 1900', in R.C. Olby, G.N. Cantor, J.R.R. Christie and M.J.S. Hodge (eds), *A Companion to the History of Modern Science* (London: Routledge, 1996), pp. 838–52. For an overview of Kuhn's position see P. Hoyningen-Huene, *Reconstructing Scientific Revolutions* (Chicago: Chicago University Press, 1993).

the irregularities in the orbit of Uranus. Scientists of that historical period tried to explain the problem by suggesting that a hitherto undiscovered planet, Neptune, was in fact producing the irregularities – and they succeeded. This is a typical example of normal science: new discoveries are made, but the paradigm is unaltered.

Many a historical example of science in its normal phase offers evidence for the fact that the perseverance shown in the face of contrary evidence is considered meritorious, and not an instance of scientific misconduct, as a strict defender of falsificationism would say. Perseverance, however, has its limits. What happens when a particular puzzle becomes really hard to solve? Philosophers such as Norwood Russell Hanson, Paul Feyerabend and Thomas Kuhn have shown that such situations usually give rise to an extended investigation in the area of the anomaly. The aim is to try to save the contradicted theory at all costs. When such endeavours fail, the real crisis sets in. Doubts about the validity of the dominating theory give rise to doubts about the validity of the entire paradigm. Younger scientists start losing faith in the old ways of doing things. They become suspicious of their old text-books. They start seeking new styles of doing science. They may come up with the solution of the anomaly by moving into a totally new, self-consistent way of seeing things.

The crucial novelty in this account of scientific development is the gap that divides the old paradigm from the new. The paradigm is said to carry *within itself* all the rules that determine not only which theories are acceptable but also which kind of observations are relevant. As regards justification of knowledge claims, a paradigm is self-contained. This holds both for the old paradigm and for the new. The consequence is that the move from one paradigm to another can never be the result of rational deliberations by scientists in their old paradigm. The argumentative resources for the old-fashioned scientists will lie within the old paradigm, while the argumentative resources for the break-away, younger scientists will lie within the new paradigm. If reasoned argumentation, in this radical account of scientific development, does not count anymore, one may legitimately ask: what counts then? The only response that makes sense seems to be the claim that a paradigm-shift is somewhat closer to an experience of religious conversion than it is to a rational deliberation. The popularity of the new paradigm will first have only a few bold adherents. As the benefits of escaping the original anomaly become more evident, more and more of the old-paradigm scientists will be converted. The old paradigm will therefore fade away not because it is intrinsically erroneous. The very criteria of right and wrong are situated within the paradigm itself. It will fade away simply because it loses adherents. Any stubborn, old-fashioned scientist who wants to remain in the old paradigm when the majority of other scientists have moved ahead will, by that very fact, cease to be a scientist.

The effects of a scientific revolution go very deep. Kuhn writes: 'after a revolution scientists work in a different world'.[2] The very nature of rationality undergoes a change. When younger scientists start losing faith in the old ways of proceeding, they abandon everything. They cannot argue rationally for the benefits of adopting

2 Kuhn, *The Structure of Scientific Revolutions*, p. 135.

the new paradigm. Social, political and cultural forces come into play. Moreover, one should include the effects of strong personalities in scientific circles: individuals who determine funding policies for future projects. It is the combined effect of these forces that somehow pushes the old paradigm out of existence. No rational decision is possible. No crucial experiment can help either, because the very determination of which observations are relevant and which not is part of the debate. According to this account, therefore, the very definition of rationality changes repeatedly in the course of history. What was considered true for scientists of, say, the seventeenth century, was true for them. What is considered true by us today, is true for us. There can be no comparison. There is no scientific progress in the sense of getting closer and closer to the truth. Our major scientific theories are changeable pictures of the world. They stand next to each other not in a sequence of greater and greater truth-content, but in no sequence at all, somewhat like pictures in an art gallery.

Admittedly, Kuhn did not simply write off all possibility of reasoning during the period of abnormal science. He conceded that there are reasons being used by contenders of different paradigms. He insisted however that such reasons function as values that can be applied differently by different people. One may, for instance, fix the fruitfulness of a hypothesis as a good reason for its acceptance; but there may still be a variety of ways how to assess such fruitfulness. He writes that, in such situations, reasons function as values that can

> be differently applied, individually and collectively, by men who concur in honoring them. If two men disagree, for example, about the relative fruitfulness of their theories, or if they agree about that but disagree about the relative importance of fruitfulness and, say, scope in reaching a choice, neither can be convicted of a mistake. Nor is either being unscientific. There is no neutral algorithm for theory-choice, no systematic decision procedure which, properly applied, must lead each individual in the group to the same decision.[3]

The situation here suggests that some general principles are available but they cannot be used to determine uniquely a way of proceeding in a concrete particular situation of theory choice. One may start seeing here a similarity here between Kuhn's analysis and the traditional concept of practical reasoning.

The radical nature of such an account of change in world-view is often expressed in two ways. First, the idea of meaning-incommensurability as regards central theoretical concepts: the power of a paradigm engenders such a high degree of meaning-holism that the meaning of a single concept depends on the whole set of concepts within which it is embedded. For instance, the term 'mass' within the context of relativity physics would, if the radical account were correct, be totally different from what Isaac Newton had in mind when he used the same word. To understand Newton's full meaning of 'mass' we have to understand not only the scientific context in which he used it, but also the metaphysical presuppositions underpinning such a context. And to understand these presuppositions, we have to understand the cultural and social world within which they were expressed and

3 Ibid., p. 200.

referred to. In short, to understand Newton's word 'mass', we have to grasp his entire paradigm, which is impossible.

The radical nature of historical change is also expressed in another way. This involves the so-called pessimistic induction, a version of the classical sceptical argument from previous mistakes. Philosophers like Larry Laudan have unearthed many an embarrassing mistake made by previous generations of scientists. The best example remains the one involving phlogiston.[4] Where modern chemists see a gain of oxygen, the pre-Lavoisier chemists used to see the exact opposite: a loss of phlogiston. Hence the old theoretical entity was not even an approximation of some entity in the new theory. The old idea of phlogiston was just a mistake. Pre-Lavoisier chemists thought they were referring to something, but in fact they were not, and there is no way we can save their judgement. A radical account of scientific revolution leads to the following kind of pessimism. If scientists have been radically mistaken as regards their theoretical entities in the past, they may very well be mistaken about the theoretical entities we accept today. This is an inductive argument. The occurrence of many mistakes in the past indicates that a mistake is probably being made now.

Let me sum up this first step of my argument. According to this kind of radical understanding of scientific revolutions, it is not difficult to see how personal value judgements become inevitable. When a new paradigm becomes available, rationality itself is shifting. The only criteria available for proceeding are personal ones related to the particular dispositions of the individuals involved.

At this point, one can foresee an objection in the form of a question: isn't this radical account totally incapable of explaining the impressive success of science? Addressing this question constitutes the second step of my argument. Richard Boyd has offered some convincing arguments to show that the best explanation of the instrumental and predictive success of mature scientific theories is that these theories are approximately true.[5] Scientists devise experimental set-ups, form their expectations, and choose among competing theories always in reference to their accepted background theories. These theory laden procedures result in correct predictions. The best way to explain this success is to admit that the central theoretical statements in the background theories are approximately true. Put simply, today's predictive success adds justification to yesterday's truth-claims. The process is cumulative.

---

4 For Laudan's plethora of non-existent theoretical entities see his 'A confutation of convergent realism'. For historical details about the particular case of phlogiston, see Frederic L. Holmes, *Lavoisier and the Chemistry of life: an exploration of scientific creativity* (Wisconson: University of Wisconsin Press, 1985).

5 Richard Boyd, 'Scientific Realism and Naturalistic Epistemology', in P.D. Asquith and T. Nickles (eds), *PSA 1980*, vol. 2 (East Lansing, MI: Philosophy of Science Association, 1981), pp. 613–62; 'Realism, Approximate Truth and Philosophical Method', in C.W. Savage (ed.), *Scientific Theories* (Minneapolis: University of Minnesota Press, 1990), pp. 355–91.

Is this, as some have argued, viciously-circular reasoning?[6] It apparently employs the very type of argument whose cogency is under discussion. The initial theory-laden theories are arrived at by accepting the validity of inference to the best explanation. These theories are then seen as successful. They are therefore considered approximately true. The overall conclusion seems to be that inference to the best explanation is valid. This however was already assumed in the first instance. Therefore there seems to be some circularity here. Without going into a detailed evaluation of such debates, we can see that the charge of circularity is somewhat contrived.[7] What the Boyd argument is starting with is not a claim that inference to the best explanation is valid. He does not need certainty on this point. All that he requires is that inference to the best explanation is not clearly unreliable. Saying that it's not clearly unreliable is not the same as saying that it's clearly reliable. There is a space between these two claims. It is precisely this space that this entire debate is about. Theory-laden procedures are devised by the tentative use of inference to the best explanation. The success of this operation corroborates the entire operation. It shows both that the theories are approximately true, and that we were justified in considering inference to the best explanation as not unreliable. The circularity is not vicious.

The radical account of scientific revolutions needs therefore some refinement. This call for refinement comes from the obvious empirical fact of the success of science. Another call for refinement comes from logical considerations alone. Pure incommensuarability between conceptual schemes is impossible. On this point, Donald Davidson has supplied us with a kind of *reductio* argument.[8] The very idea of a group of beings in general, or of scientists in particular, who think and reason in accordance with thought-constraints completely different from our own leads to an absurdity. Think of what is involved in the very positing of this situation. There are two steps. We can recognise that we are in a situation as the one described, first of all, when we judge that the others, in the other paradigm, had been capable of thinking or of reasoning. Then we conclude that their thinking and reasoning is completely different from our own. The inherent problem should be evident now. At the first step, how can we identify a process and call it thinking if it is completely different from our own? For thinking to be recognised as thinking at all, it must have some common element with *our* thinking. The constraints we have on thinking partly define what counts as thinking.

Another way of making the same point is to refer to the well-known idea attributed to Willard van Orman Quine: any statement can be held true if we make

6 For instance, Arthur Fine, 'Piecemeal Realism', *Philosophical Studies*, 61 (1991): 79–96.

7 A good overview of the debate can be found in: Stathis Psillos, *Scientific Realism: How Science tracks the Truth* (London: Routledge, 1999), chapter 4.

8 D. Davidson, 'On the very Idea of a Conceptual Scheme', *Proceedings and Addresses of the American Philosophical Association*, 57 (1974): 5–20; reprinted in D. Davidson, *Inquiries into Truth and Interpretation* (Oxford: Clarendon Press, 1984), pp. 183–98.

sufficiently drastic changes elsewhere in the system. Following Quine, a defender of radical incommensurability may insist that we can indeed accept that scientists in other paradigms had been in fact thinking even though such a process is completely different from what we call thinking. To accept this, so the objection goes, all we need is to make the right adjustments within our presuppositions. This use of Quine, however, is misleading. The statement under consideration in this case is not an isolated one. It involves what we take to be thinking. To accept that scientists in other paradigms are thinking completely differently from us, we need to make not only drastic changes *within* the system – that would not be enough. We would also have to renounce one of the main points of having such a system at all. We would have to renounce the primordial attempt to clarify the world to ourselves in any intellectually satisfying way. And if *that* is allowed, then anything goes.[9] So in the end we always arrive at an absurdity. The *reductio* argument is complete. We are entitled therefore to hold that there is necessarily a hard core of paradigm-independent criteria of rationality. There is no way of avoiding the need to refine the radical account of scientific revolutions.

The relevance of these arguments should now be clear. In my reasoning so far, I have taken two steps. I started with the idea that a radical understanding of scientific revolutions, makes personal value judgements in science inevitable. The second step was essentially a critique of the radical understanding of scientific revolutions. I showed that some refinement is necessary. The two reasons I gave for this claim are, first, the impressive success of scientific explanation, and second, the logical impossibility of having two different but recognisable rationalities. A less radical account of revolutionary periods in the history of science is needed.

The third step in my reasoning will therefore be to explore whether a less radical account of revolutionary periods in the history of science would do away with personal value judgements. As has been proposed by Imre Lakatos, such a less radical account should avoid the idea of paradigm, because the identity-conditions of a paradigm are not readily available. The suggestion is to use the less vague concept of scientific research programme.[10] It can be seen as having a hard core of

9 The point about the quotation from Quine is made in E.J. Craig, 'The Problem of Necessary Truth' in S. Blackburn (ed.), *Meaning, Reference and Necessity* (Cambridge: Cambridge University Press, 1975), pp. 1–31. For a historical overview of holism in philosophy of science see: A.A. Brenner, 'Holism a Century Ago: The Elaboration of Duhem's Thesis', *Synthese*, 83 (1990): 325–35. Fundamental premises associated with thinking have had various formulations. Kant calls it 'the transcendental unity of apperception', signifying thereby the fact that all of the representations I have are necessarily mine. For Strawson, the essential point is that it must be possible for someone to distinguish between states of himself and states not of himself. For Ross Harrison, in any comprehensible world there must be reasons available to the protagonist enabling him to distinguish between those judgements of his which are true and those judgements of his which are false. A valuable source of information about these issues is R. Harrison, *On what there must be* (Oxford: Clarendon Press, 1974).

10 Imre Lakatos, 'Falsification and Methodology of Scientific Research Programmes,' in I. Lakatos and A. Musgrave (eds), *Criticism and Growth of Knowledge* (Cambridge:

theories that are considered indisputable by the researchers of that programme, and surrounded by a protective belt of auxiliary hypotheses that explain why the hard core theories should not be rejected when contradictory evidence arises. The task of scientists working in a given research programme is therefore to formulate the simplest, and thus most acceptable, auxiliary hypothesis to save the hard core, to test such formulations by experiment, and when these hypotheses are inadequate, to adapt them or substitute them so as to ensure the credibility of the hard core.

On this account, the history of science is seen as a study of how the hard core of research programmes fares in empirical predictions. A hard-core can be self-sufficient and capable of explaining all the available observations. Otherwise, it can be weak, in the sense of needing a considerable amount of ingenious and contrived auxiliary hypotheses to keep itself afloat in the face of contrary evidence. Hence the way a research programme copes with new observations is a valuable measure of its acceptability. This feature allows a research programme to be called either progressive or regressive. Within a research programme, theories may shift from T_1 to T_2, and so on. A theory-shift is progressive when three conditions are met. It is progressive when T_n accounts for previous successes of T_{n-1}, when T_n has greater empirical content than T_{n-1}, and when some of the excess content of T_n has been corroborated.[11] It is regresssive when the auxiliary hypotheses needed to protect the hard core become more and more numerous and intricate, and if the hard core produces only a few new, empirical predictions. In this way, the research programme loses credibility and is finally abandoned.

This account of historical development of science has one obvious advantage over Kuhn's paradigm-approach. It restores the crucial role of rational deliberation. To decide whether a given research programme is progressive or regressive, one isn't reduced to sitting down and waiting to see what the social forces will decide. Reasoning cannot be abandoned. For this decision, one needs to determine whether the auxiliary hypotheses fabricated to protect the hard core are few in number and easy to formulate, and whether the programme as a whole is making new, verifiable empirical observations. This fact, and only this fact, is relevant to the question about the progressiveness or regressiveness of the research programme.

Saying that the history of science involves rational deliberation does not imply that rational deliberation determines all that is done in scientific development. A certain role for personal value judgement cannot be eradicated. Lakatos allows us to distinguish very clearly between, on the one hand, the methodological appraisal of a research programme, and, on the other hand, the decision on the part of the individual scientist whether to continue applying it. He writes: 'It is perfectly rational to play a risky game: what is irrational is to deceive oneself about the risk.'[12] Rationality is involved in acknowledging the situation as risky. It is involved

Cambridge University Press, 1970).

11 Ibid., pp. 116–8.

12 Imre Lakatos, 'History of Science and its Rational Reconstructions' in R. Buck and R. Cohen (eds), *Boston Studies in the Philosophy of Science*, vol. 8 (Dordrecht: D. Reidel, 1971),

in making a judgement about the risk. As regards all this, there is an element of objectivity. There are criteria that allow convergence of opinion when various protagonists are involved. However, when it happens that the individual scientist is to choose whether to continue investing time and energy in that particular research programme, personal value judgement is all that is left as a guide.

To recapitulate, therefore: I have argued that in the very radicality of the paradigm-account lies its weakness. It certainly gives an important role to personal value judgements in the development of science but is unable to explain the success of science. The refined account in terms of research-programmes is richer. It sees the development of science as a mutable agglomeration of research programmes living side by side, as it were, in the course of history. Some are progressive and survive. Some are regressive and degenerate and disappear. Unlike the radical account, history is not seen as a series of identifiable temporal parts succeeding one another in a neat way. What we have is a combination of overlapping programs determining an overall continuity of the practice we loosely call science, even though some are initiated and others are seen to be degenerate and are abandoned. The continuity involved here is similar to that of a thread which is continuous even though it is made up of overlapping fibres, very few of which may run through it all. The major point of interest for my purposes is the interplay between rationality and personal value judgements in this refined account of science. Logically explicit methodological rules are at work all the time, but the role of personal dispositions cannot be eradicated from the process. Ian Hacking has summarised this point as follows:

> As for the future, there are few pointers to be derived from his [Lakatos'] 'methodology'. He says that we should be modest in our hopes for our own projects because rival programmes may turn out to have the last word. There is a place for pig-headedness when one's programme is going through a bad patch. The mottos are to be proliferation of theories, leniency in evaluation, and honest 'score-keeping' to see which programme is producing results and meeting new challenges.[13]

Personal attitudes, virtues and vices are not foreign to the way science works. There is an element of practical wisdom that goes beyond ratiocination.

The line of reasoning up to now has been divided into three steps. The first was an exploration of the most radical account of scientific revolutions; the second was about the weaknesses of this account; the third was about an improved refined version. The original overall aim was to defend both the claim that there are genuine, radical shifts in the history of science and the related claim that these shifts inevitably involve personal value judgements. The fourth and final step now is to explore the interface between theoretical and practical wisdom, within the context of scientific development. In this way, I hope, a clearer picture will become available of how science and personal attributes are mutually dependent.

---

p. 104n.

13 Ian Hacking, *Representing and Intervening* (Cambridge: Cambridge University Press, 1983), p. 121.

That there are some very significant and radical changes in scientific world-view cannot be denied. How best to describe them is another question. If we do it in terms of paradigms, we end up in trouble. If we do it in terms of research programmes, we have less trouble. The gap between the old science and the new science becomes somewhat more understandable. Nevertheless, when impressive cases of deep change within a theory, like the one involving phlogiston, are accounted for in terms of research programmes, one cannot deny a certain residue of incommensurability. Perhaps the best way to understand this gap is the one recently suggested by Bas van Fraassen. This is in terms of asymmetric intelligibility. For him, the gap can be seen as

> characterised by a remarkable historical asymmetry. From the posterior point of view, the prior can be made intelligible and the change ratified. From the prior position, however, the posterior view was absurd and the transition to it possible but incapable of justification.[14]

What we have here, therefore, is a change in world-view that shows retrospective intelligibility of the old view on the part of the new generation of scientists, together with a prospective unintelligibility of the new view by the passing generation. There is a jump into the darkness, a leap of faith that brings new light on the whole. One can understand why Kuhn used the term conversion to describe what happens during a paradigm-change.

This way of acknowledging radical changes of world-view highlights the fact that an epistemological risk is involved. As we know even in our day-to-day business, risk is very much a part of life. If we wanted to avoid all risk of making mistakes, we could just renounce having opinions and believe tautologies only. Such a move would ensure certainty as regards our beliefs, but the information content of what we actually believe would be drastically reduced. Avoiding at all costs the risk of error would leave us with very little truths. This is exaggerating in one direction. If we exaggerate in the opposite direction, it would be just as bad. Suppose we want to avoid the risk of remaining with zero, or near zero, truth-content. To attain this, we could just believe everything that comes to mind. This attitude would certainly allow us to believe much that is true, but it would also reduce the accuracy of our claims. We could be sure that the set of all our beliefs would include many that are true but it would also include many that are not. There would be no security in what we claim. It is clear therefore that what we need is a healthy equilibrium between these two extreme attitudes. Epistemological risk is unavoidable. We need to learn how to live with it: how to arrive at the right amount of risk that makes life flourish well. This is where an element of practical wisdom over and above ratiocination comes in. In van Fraassen's words:

14 Bas C. van Fraassen, *The Empirical Stance* (New Haven: Yale University Press, 2002), p. 65.

> The telos should be the search for a properly or well balanced body of opinion, with some equilibrium of information and security ... Now although truth and error are objective categories, handed to us by nature itself, so to speak, the measure of balance is not![15]

So, in a nutshell, what I have been advancing is an argument that results in the awareness that the refined account of radical shifts in terms of scientific research programmes has not eliminated all elements of personal judgements. These elements start to look quite similar to the classical 'practical wisdom' involved in moral decisions.

Consider for instance Hilary Putnam's suggestion that, during a period of Kuhnian abnormal science, the scientist needs skills that go beyond theoretical description.[16] He sees this as a parallel between knowledge in the natural sciences and knowledge in the social sciences. Both depend to some extent on unformalised practical knowledge. Both need skills that are too complex to describe fully by a theory. The specificity of the decision-situation defies generalisations. It is arguable that even in basic everyday environments such skills are at work. The ability to use psychological predicates correctly implies a skill. Putnam reminds us that it couldn't be a theory because, if we formulate a theory of the form 'Person A is jealous if and only if behaviour pattern B is observed', we need to add 'unless special circumstances obtain'. This is typical of the practical syllogism: its dependence on the *ceteris paribus* clause. Further considerations relevant to this point are well rehearsed in Richard Bernstein's *Beyond Objectivism and Relativism: Science, Hermeneutics, Praxis*, where we find the following illuminating recognition:

> One must be sensitive to and acknowledge the important differences between the nature of scientific knowledge and other forms of knowledge, but the more closely we examine the nature of this scientific knowledge that has become the paradigm of theoretical knowledge, the more we realise that the character of rationality in the sciences, especially in matters of theory-choice, is closer to those features of rationality that have been characteristic of the tradition of practical philosophy than to many of the modern images of what is supposed to be the character of genuine *episteme*.[17]

An algorithm able to dictate rational, unanimous choice between theories is impossible. The criteria of choice function as values that influence choice not as fixed rules that makes the choice inevitable. Accuracy, consistency, scope, simplicity, fruitfulness are such values not rules. They are general principles and the particularity of the situation defies a deductive strategy.[18] It is easy to understand why, at this

15 Ibid., p. 87.

16 H. Putnam, *Meaning and the Moral Sciences* (London: Routledge & Kegan Paul 1978), pp. 71–72.

17 Richard J. Bernstein, *Beyond Objectivism and Relativism: Science, Hermeneutics, Praxis* (Philadelphia, University of Pennsylvania Press, 1991), p. 47.

18 These basic ideas are spelt out clearly in T. Kuhn, 'Objectivity, Value Judgement and Theory Choice', in *The Essential Tension: Selected Studies in Scientific Tradition can Change* (Chicago: University of Chicago Press, 1977).

point, Bernstein applies Aristotle's idea of *phronesis* to account for the way science develops during a radical shift. On the nature of *phronesis* there has been much discussion: the element I am highlighting here is the lack of deductive precision involved in such reasoning. It is certainly the case that one individual scientist who has shown his skill at choosing the right way through the darkness of a radical shift in scientific world-view doesn't necessarily have to be a good person in the moral sense. Nevertheless, the moving from general principles to decisions in particular situations is analogous in both cases.

Consider, for instance, the celebrated case of Galileo Galilei. Was there a role of practical wisdom in this case? Let me recall the major details first. One of Galileo's major contributions was his claim that the Copernican, heliocentric system should replace the Ptolemaic, geocentric one not only because the former facilitated calculation when compared to the latter, but also because the former was a true picture of how things are. Galileo's main argument to justify the first claim was convincing. He showed that the various Ptolemaic equivalent systems functioned well, but only to a certain extent. They could not explain new observations made available through his telescope. The Ptolemeic systems were not saving all the appearances, while the Copernican system was. That was enough justification for his first claim. His second claim was more problematic. He needed an argument to show that the heliocentric system was not just one descriptive device among possible others, but the one and only correct description. If, during his time, there had not been any other mathematical model that explained the newly observed phenomena, he could have easily succeeded in justifying this second claim. Tycho Brahe, however, had proposed a model wherein the Sun was not only revolving around a stationary, fixed Earth but was also the centre of orbit of all the other planets, which were thus dragged around the fixed Earth as they go around the sun. This model saved all the appearances. It saved the planetary retrograde motion Ptolemy had explained. It saved the newly observed phenomena that Copernicus was explaining. It also saved the everyday appearance that the Earth was comfortably stable and not whirling around in space at terrifying speeds – a point much more important in sixteenth-century public consciousness than it is today. So, as regards saving the appearances, Galileo was faced with the old problem of equivalent systems.[19] Both parties in the debate were aware of the fact that one sure way of distinguishing the Copernican system from Brahe's system was the observation of stellar parallax. This effect, however, is so small that it took astronomers about two hundred years before telescopes strong enough to observe it were constructed.[20] Galileo remained without his most precious piece of evidence till he died. Some Church scholars, like Cardinal

19 Brahe's system is not observationally equivalent to the Copernican one; the former predicts full phase-changes of all planets while the latter predicts full phase-changes for Mercury and Venus only. This point was not considered important in the time of Galileo because it was probably believed that the observation of the phase-change of planets other than Mercury and Venus was simply a matter of developing more powerful telescopes.

20 First observed by F.W. Bessel in 1838.

Robert Bellarmine, advocated an attitude of caution until more definite evidence is available.

This familiar, crucial event in the development of science has been studied from various perspectives. What I want to highlight here is the way it illustrates the element of inevitable risk involved in such fundamental shifts in scientific world-views. On a superficial reading, the episode may look like a case of obscurantism. On more detailed analysis, however, it becomes clear that neither Galileo nor his opponents had definite proof for their respective positions. The more data we bring into our understanding of the cultural context of the late sixteenth and early seventeenth centuries, the more we appreciate the critical balance in which the protagonists of that debate had to deliberate. Arguments and counter-arguments in support of rival theories and philosophies were not conclusive. Galileo's Aristotelian opponents were certainly not at the end of their tethers. There was space for reasoned argument on both sides. It is only in retrospect that we now acclaim Galileo as the true heir of past insights and aspirations concerning a secure method of describing the nature of things. At that time, what was needed was a kind of wisdom that allows advancement even though reasoned argument seemed to stall to a halt. The practical wisdom needed here involved the decision on how to take the right amount of risk that makes life flourish well. The Galileo's case, therefore, is a good example of the two major points made above. First, it illustrates how the radicality of some shifts in world-view is real and can be understood in terms of the asymmetry between prospective unintelligibility and retrospective intelligibility. Second, it illustrates how ratiocination must give place to personal value judgements.

Let us count our gains. At the beginning of this chapter, I set myself the task of defending the thesis that there is no algorithm for choosing the correct way forward during major upheavals in scientific thinking. Arranging my reasoning in four steps, I drew the conclusion that the way forward in such upheavals will inevitably involve personal value judgements. The set of these personal value judgements constitute a certain kind of skill or practical wisdom. This conclusion essentially means that personal attitudes and dispositions affect some crucial periods of the history of science. It is time now to move on to explore the interaction between personal value-judgements and science in the other direction. It is time to explore how the historical awareness introduced in science affects personal attributes.

Evaluation of Theories and Growth in Virtue

It is useful to recall the initial assumptions of this kind of investigation. Firstly, I am assuming that the scientific mentality is a complex agglomeration of cultural trends that is affected to some extent by the intricacies of the logic of scientific discovery, as mentioned in Chapters 2 and 3. It is affected also by the nature of scientific explanation, as discussed in Chapter 4. Here, I am assuming that the scientific mentality is affected also by the particular questions arising from the historiography of science. Especially in the wake of the last decades, when such questions have established themselves as a fundamental part of the science-studies curriculum, the scientific mentality has undergone a kind of transformation. The doubts concerning the canon of rationality have left an indelible mark on the way scientists conduct their research and on the way they try to understand the research of previous generations. I will be referring to this dimension of the scientific mentality by the term historical consciousness.

Secondly, I am assuming also that this particular aspect of the scientific mentality has an even deeper effect: an effect on the dispositions of the individual scientist as a person. Such an effect, over a long period of time, either enhances or inhibits specific personal attributes of that individual. These character traits are virtues when they help the person flourish or live well. As in the discussion of previous chapters, I am not pretending that the concept of human flourishing is straightforward or unproblematic. I feel entitled to use it, however, because it is not more obscure than the concept of rationality or happiness. Such concepts are indispensable in various philosophical discussions, and to decide not to use them because they cannot be fully analysed would certainly be unrealistic. My discussion here will therefore proceed in a style reminiscent of virtue ethics, with, however, a significant difference. Ethicists are normally, and understandably, concerned with actions. They attempt to discover which action is to be done, for instance: is abortion right? This can be done using various ethical theories, including that of virtue.[21] I am engaging in a different kind of reflection. I am interested in the influence a specific way of living and thinking may have on character traits. The central question for me will be: does this specific cultural element, in this case the historical consciousness within the scientific mentality, make the person become more virtuous? My focus, therefore, is not on any one action in particular, but on a person's development. Will he or she become more capable of doing acts that are in line with human flourishing?

Having secured this basic framework for my strategy, I will now proceed as in previous chapters. I will start with an inquiry into the two major negative effects the historical consciousness may have on personal dispositions. These two negative effects will be shown to be associated with opposite vices. Following the Aristotelian insight, I will then attempt to describe a particular virtue by seeking the mean between the two extremes.

I start with a point often highlighted in the first section of this chapter: the issue of incommensurability between historical paradigms. It was shown that taking the historical dimension of science seriously can lead to the awareness that a certain degree of incompatibility between the various periods is inevitable. This element of incommensurability, or distance, can be expressed in a radical way by claiming that natural science is only an exercise in saving the appearances and nothing more. It can also be expressed by endorsing the thesis of scientific relativism. What is significant in the present argument is to ask what effect this awareness may have on the personal dispositions of the individual scientist. Since the issue here has to do with the long

---

21 A good example showing virtue-ethics 'in action' is Rosalind Hursthouse, 'Virtue Theory and Abortion', *Philosophy and Pubic Affairs*, 20 (1991): 223–46.

chain of knowledge-acquisition linking one generation to the next, it is plausible to suggest that the effects such beliefs will have in the long run on the individual include a certain distrust as regards the validity of 'received views'. It is certainly the case that a certain amount of systematic scepticism forms part of the very fabric of scientific practice. The rigorous scientist is looked upon as the one who doesn't often take claims for granted. This however does not mean that literally every element of knowledge in any particular research programme needs to be verified by each generation. Much depends on trust. And this is the personal disposition that suffers if the historical consciousness is appropriated in its negative aspect. An over-emphasis on incommensurability will tend to make a person distrust tradition.

Is there a link between this effect of affirming radical incommensurability and the idea of human flourishing? Does one become more prone to do things that go against human flourishing if one endorses radical incommensurability? One way to reply to this question is to start from the conclusion already drawn in the previous paragraph, namely that a person who accepts radical incommensurability becomes, by that fact alone, sceptical as regards all kinds of received views. An attitude will be reinforced whereby the person harbours doubts about the value of the work done by previous generations. As such, this attitude does not appear straightforwardly detrimental to human happiness. If the individual develops an attitude of special care to avoid mistakes, there is certainly nothing wrong. The case under discussion, however, is an extreme one. It is one in which the individual develops an exaggerated distrust of the very idea of truth. And this leads to an undesirable situation because, as a result, the individual will become open to repeating the mistakes of previous generations. The idea of learning from previous generations is a natural extension of the parent-child relationship, and can be called a basic element of human flourishing. Any attitude that undercuts such learning is detrimental to the individual. From here, it is not difficult to see that accepting radical incommensurability produces a vice: a 'negative' disposition in the individual.

One common criticism to this line of reasoning is that there are some effects resulting from affirming radical incommensurability that seem to be quite desirable. If true, this would mean that such an affirmation would be completely in line with human flourishing, and therefore certainly not linked to any vice. For instance, it has been argued at some length that the following three results of belief in radical incommensurability are definitely positive.[22] Firstly, such a belief will ensure the free development of the individual; secondly, it will ensure possible, hitherto unimagined, steps of scientific progress achievable only through counter-inductive thinking; and thirdly, such a belief will ensure the avoidance of the chauvinism of the natural sciences. Such effects, however, are only apparently positive. Not much reflection is needed to uncover an underlying problem in all three points. Free development of the individual is fine, but only if the term 'free' is correctly understood. Whenever the acceptance of radical incommensurability results in a free individual in the sense

22 I am taking as my example Paul Feyerabend, *Against Method* (London: Verso, 1993).

that he or she is more prone to repeat yesterday's mistakes, that can hardly be called a positive result. Freedom from mistakes is certainly preferable to freedom to repeat them. As regards the second and third points, a similar strategy will easily unmask the problem. The occasional need for counter-induction and the problem of scientific bigotry are two factors resulting from a certain distorting effect the past may have on the present. Being aware of such distortion is fine, but only if the right kind of distortion is being considered. Counter-induction makes sense when there is an accumulated body of knowledge-claims to contrast against. It is occasionally useful to enhance creativity. If, however, the belief in radical incommensurability makes a person engage *exclusively* in counter-induction, he or she will end up thinking that the past is one long mistake enslaving the present. The vulnerability to fall into yesterday's mistakes will resurface. The same line of argument holds for the drive to counteract the worrying chauvinism of science. Anything that leads to a personal disposition that makes the individual more prone to repeat yesterday's mistakes cannot be considered in line with genuine human flourishing.

At this stage of my inquiry, therefore, it is reasonable to conclude that one particular aspect of the historical consciousness discussed in this chapter leads to a vice. It has been shown that affirming radical incommensurability leads an individual to act in a way that goes against genuine human flourishing. I will now proceed by exploring an opposite tendency that can result from the same historical consciousness. Just as some philosophers tend to defend radical incommensurability, others, on the contrary, tend to defend the exact opposite. When their awareness of history of science is heightened, they tend to consider this history an opportunity to uncover the hidden rational process in operation at every stage of development, however minute. For instance, the historical consciousness has made people like Karl Popper believe that it is possible to arrive at a fixed set of rules defining the method of natural science once and for all. Linked to this idea of a fixed, discoverable list of rules of method, is the idea that, once these rules are formulated, the discipline is fixed for all time in its essential characteristics. The only task left for scientists would be to apply this method in areas where it has not yet been tried out. It may be helpful here to compare this way of understanding science with the statute-law approach in the philosophy of law. The term 'statute-law' refers to the idea that a set of rules are instituted with the aim of biding all future cases. In this way, the community in which such legislation occurs may be said to found or create its identity, partly or wholly as the case may be, precisely by the establishment of such laws. The case of natural science, as discussed above, is the same. The rules of method, once formulated, would be constitutive of the identity of the discipline and would be binding for all scientists in all future scientific problems.

Which behaviour patterns will be enhanced in the individual because of the acceptance that the intrinsic nature of science is like the statue-law model? The basic tendency will be one that pushes the person towards over-emphasising ratiocination, over-emphasising that particular manner of logical reasoning in which each step is evident. In other words, the person will acquire a disposition to value only those actions that are manifestly prescribed by a pre-established set of rules. Any action that is spontaneous will be systematically considered suspicious and better avoided. The reason behind this is that the person will grow in the habit of acting always according to what is explicitly prescribed by known rules. In such a way of life, no space will be left for the intuitive grasp of the moral significance of an action. There will be no space for virtue as such. In fact, taken to the extreme, this position will imply that machines are the best agents.

Now the proposal I would like to defend is that this personal tendency, resulting ultimately from taking some aspects of the historical consciousness to heart, is tantamount to a vice. In a nutshell, my argument is that the disposition under consideration leaves no space for moral wisdom. As many moral philosophers ever since Aristotle have pointed out, there is a basic role played by prudence in moral philosophy. Aristotle draws our attention to the fact that we need to postulate the existence of a specific human ability to explain how a person sees the moral significance of the act he or she is engaged in. Prudence, traditionally classified as a virtue of the practical intellect, corresponds to this ability to direct a set of actions in such a way that they will fit into a fulfilled human life. Prudence, as a moral concept, goes beyond mere success. It is not the person's ability, say, of doing a scientific experiment successfully, but to see how doing it successfully fits into the overall flow of the person's life, guiding it towards happiness and fulfilment. It is clear, therefore, that, if a person is in a state where action is done only according to pre-established rules, there is an explicit rejection of the role of prudence in that person's life. The individual becomes morally blind.

An objection can be raised at this point. An opponent may point out that the jump from ratiocination in scientific reasoning to ratiocination in normal everyday deliberation and action has been too quick. In other words, individuals may very well live lives that proceed according to various standards. One may be completely rule-driven as regards scientific action while, at the same time, being open to moral intuition as regards other areas of action 'outside the laboratory'. This objection, however, is somewhat contrived. It depends entirely on a fictional clear-cut distinction between the life of science and ordinary life. Explanations we come up with and use in ordinary life conform in large part to the style of scientific procedure. This has been evident to many investigators, some of whom have even gone to the extent of calling science a refined or critical form of common sense.[23] The idea that life involves the simultaneous existence of various worlds that are completely detached from each other does not do justice to the simple fact that it is one and the same person who is commuting to and fro between these environments. Any deep effects left on the character of the person by one world will be there present in the other. The objection, therefore, can only be taken seriously in an imagined world inhabited by people who

23 For instance, C.S. Peirce, 'Pragmaticism and Critical Common-Sensism' in C. Hartshorn, P. Weiss and A. Burks (eds), *Collected Papers*, vol. 5 (Cambridge, MA: Harvard University Press, 1931–1958), pp. 346–50; 'Consequences of Critical Common-Sensism' in Ibid., pp. 351–75. I discuss this point at some length in *Holism and the Understanding of Science*, pp. 113–6.

can, as it were, undress one personality, understood here as the set of habits, and put on another. In the real world, however, a person without personality is no person at all. Moreover, many recent studies supply irrefutable evidence for the claim that, even within strictly scientific practice, there is an essential role for the learning of skills that go beyond ratiocination.[24] Tacit knowledge, of some kind or other, plays an important role. It is a mistake, therefore, to think that the ideal scientist would be one who is aware of how each of his or her actions is in accord with some explicit rule. In any account of science, there needs to be space for intuition.

Since this major objection can be refuted, I feel secure in retaining the claim that an over-emphasis on ratiocination will probably lead to a character trait that is vicious. This vice can be taken together with its opposite described previously. The pair will enable the determination of at least one virtue, perhaps the major one, that is associated with the historical consciousness under consideration in this chapter. One vice had to do with an over-emphasis on ratiocination, the other with an over-emphasis on incommensurability. Taking the mean between these two extremes in the Aristotelian way leads to a virtue best described perhaps as a kind of heuristic courage, or boldness. It corresponds to the disposition whereby a person accepts the reality of some continuity in history and yet accepts also the reality of possible radical changes at the level of the rules of conduct. Such a virtuous person does not wilt in the face of pure novelty: the person remains open to readjust his or her categorical framework in an act of intellectual conversion if need be. In terminology reminiscent of biblical theology, one would say that this virtue corresponds to the characteristic that explains how a good person embodies the fulfilment of the law by living in the spirit.

Further insights regarding this virtue can be obtained by resorting to some areas of the philosophy of law as was done previously. Here, I will resort to the specific area of case-law. The term case-law, in general, refers to the principles and rules laid down in judicial decisions together with the generalisations based on past decisions of courts and tribunals for *particular* cases. The fundamental point about case-law is that these previous cases are looked upon as normative. They are considered the source of rules to be applied for new cases. Courts reach decisions with the knowledge that their decisions are laying down *strong indications* which will be followed in the future by later courts. This is quite different, however, from the laying down of a definite system of laws for the entire future, which would be statute-law.[25] Excessive emphasis on incommensurability between scientific paradigms leads to an attitude similar to a disregard of past cases. Excessive emphasis on ratiocination, on the

24 Original work in this area is found in Michael Polanyi, *Personal Knowledge* (London: Routledge & Kegan Paul, 1958), but many sociological studies of science can also be considered relevant here in so far as they represent an approach to science that highlights practice rather than method.

25 Statute-law and case-law approaches for scientific activity is mentioned in A.F. Sanders, *Michael Polanyi's Post-Critical Epistemology* (Amsterdam: Rodopi, 1988), pp. 138–45. This is discussed further in L. Caruana, *Holism and the Understanding of Science*, pp. 102–4.

contrary, leads to an attitude similar to an exclusive veneration for statue-law. Virtue lies in a disposition corresponding to the case-law approach whereby past experience is considered a source of education on how to face the future without renouncing the present responsibility of breaking new ground: whence the description of this virtue as heuristic courage or boldness.

To recapitulate then, this chapter has dealt primarily with the historical dimension of science, with particular attention on how theories should be evaluated. The aim was to find out how it comes about that some are accepted and some not. There are essentially two ways this distinction may arise. It is either a matter of rational deliberation on the part of scientists or it isn't. The first part of this chapter was on how the crucial shift in a scientific revolution could be understood with no, or little reference to rational deliberation. This account of science was based on how things happen in the course of history, as opposed to how things *ought* to happen according to some preconceived logical procedure. The alarming conclusion was that when there is a deep change of paradigm, the shift is caused by social forces which cannot be expressed in terms of a rational procedure. The various paradigms within which different views of the world are expressed are incommensurable, especially as regards theoretical entities and as regards claims about the deep nature of things. As an account of science, this was shown to be implausible. The reason was that the success of science as regards prediction is not a miracle. It cannot be explained except by holding that the great majority of its claims about theoretical entities are true. Hence a revised account of science was sought in terms of something that includes more than one theory and yet is not so vague as the term 'paradigm'. The revised account, in terms of research programs, was shown to be plausible on two counts. First, because it gives a specific role for rational deliberation during the substitution of one theory by another. Second, because it allows a satisfactory degree of continuity in our reading of the history of science. The final section involved a careful re-reading of this account of science with the aim of exploring how the historical consciousness that gave rise to it in the first place can affect the deep character traits of the individual involved. I concluded that an over emphasis on incommensurability between paradigms can lead to a personal disposition that deviates the person from attaining genuine happiness. Likewise, an over-emphasis on ratiocination can lead a person to negate the role of intuition in action. Using these two negative dispositions as place-holders, I argued that the virtue associated with the historical consciousness can be adequately described as heuristic courage.

It is now time to take the results of this exploration into the realm of virtue and incorporate it with what was concluded in each of the previous chapters. The basic element of the overall project is always the same, namely to understand better how the scientific mentality affects the life of the individual as regards virtue. When the individual is placed at the centre of attention, scientific activity can be seen as determining a particular way of life.

Chapter 6

Science as a Way of Life

The philosophical work accomplished in previous chapters can be described as a two-dimensional enterprise. Each chapter dealt with a major aspect of scientific practice, and then proceeded to uncover some significant effects that aspect could have on the character of the individual engaged in it. Some attributes of the scientific mentality, as it were, were shown to be associated with some attributes of the individual. The value of such an investigation lies in its highlighting the personal dimension. In science studies, this dimension is often considered beyond the scope of the discipline, the consequence being that such effects often go unnoticed.

A rapid overview of the fruits of this investigation will be helpful here. After some preliminary reflections on the nature of virtue in Chapter 1, I proceeded with an overview of philosophical problems related to observation. In Chapter 2, I drew the conclusion that the virtuous person is the one who acts within the perimeter of a right balance between the scientific and the manifest images. On the one hand, even though some of our commonly held beliefs may be mistaken, the virtuous person recognises that they constitute the inevitable point of departure for any discovery, and acts accordingly. On the other hand, even though some of our scientifically held beliefs may be mistaken, the virtuous person recognises that they constitute the results of the exercise of our intelligence, and acts accordingly. The scientific image may need constant refinement, just as much as the manifest image does. In this context, the virtuous person is the one who acknowledges the central role played by common sense and common practice, admits the possibility of error within the implications of this set of beliefs, and hopes that natural science or some other discipline will help remove prejudices and gain more true beliefs.

Chapter 3 dealt with method. The focus was mainly on induction and falsification. I defended the claim that a full understanding of both induction and falsification needs to refer to the essential link between, on the one hand, the purely logical discourse and, on the other hand, what people typically do in everyday situations. Having shown the inevitable significance of this link, I examined how the scientific method may affect the habits of the individual person. The virtues I highlighted were two: first, reflective equilibrium, manifested by acts that give due importance both to what is universal and to what is particular, and, second, prudential risk-taking. This latter virtue makes the person aware of both the strong points and weak points of our intellectual faculties.

The next aspect of the scientific mentality to be discussed was explanation. I argued that explanation is not an abstraction unrelated to personal attributes. In fact, morally significant personal attributes may be affected by explanation. The main justification for this came from the inevitable context-dependence of scientific explanation. A relevance relation is an important constituent of explanation. This relevance relation reflects the background knowledge of the community of inquires where the explanation occurs. It also reflects the ontological richness of the explanandum that allows different levels of inquiry. From here, I proceeded by identifying how genuine human flourishing may be hindered either by neglecting the context-dependence of explanation or by exaggerating it. The virtuous person is the one who gives the context-dependence of scientific explanation the right amount of importance.

Finally, in Chapter 5, I turned to the historical consciousness in science studies. I examined the familiar thesis that there is no algorithm for choosing the correct way forward during major upheavals in scientific thinking. I advanced the argument that the way forward in such upheavals will inevitably involve personal value judgements. The set of these personal value judgements constitute a certain kind of skill or practical wisdom. An over-emphasis on incommensurability between paradigms can lead to a personal disposition that makes the person wander away from attaining genuine happiness. Likewise but in a contrary way, an over-emphasis on ratiocination can lead a person to negate the role of intuition in action. I concluded that the virtue associated with the historical consciousness can be described as heuristic courage.

It is now high time to situate the entire inquiry within a broader context. Any discussion on the possible links between science and virtue is bound to touch sensitive issues dealing with the nature of philosophy itself. My analysis so far suggests that philosophy is not a kind of natural science. It suggests rather that philosophy is a discipline dealing with how to live well. The first section of this concluding chapter will therefore deal with philosophy not as a means of attaining a correct representation of the world but as a means of attaining a good way of life. I will then apply the general ideas explored in this first section to the more specific case of natural science. The aim in the second section will be to determine how natural science itself can be seen as a way of life: not only a source of knowledge but also a kind of wisdom. In the third and final section, I will offer some reflections on possible repercussions on the much debated area of the interaction between science and theology.

Philosophy and Life

Many are familiar with the classic definition of philosophy as the love of wisdom. As etymology goes, this is quite accurate. It says little, however, about how the discipline has actually been perceived through the centuries. Let me start by examining a particular dimension of philosophy that has received special attention in recent scholarship, namely the sapiential or existential dimension: philosophy not as a discipline that leads to true representations of the world, but as one that helps people live good lives. I will argue that seeing philosophy in this way, seeing it as a way of life, is entirely plausible, and even crucial for our own times. To do this, I will mention and evaluate a series of reasons for this claim that occur in recent literature. The aim of highlighting this dimension of philosophy is to be in a good position to discuss in the following section the particular case of natural science.

A recent protagonist in the re-evaluation of what it means to philosophise is Pierre Hadot, whose book *Philosophy as a Way of Life: spiritual exercises from Socrates to Foucault* is the culminating effort of his relentless search for the more subtle meanings of philosophy in various texts ranging from the times of Ancient Greece to the present.[1] His initial, basic discovery is that the intention of the philosophers of classical antiquity was in the first instance to form people, in the sense of educating them to live well. They were not concerned with conveying information but with helping readers undergo a conversion, a total transformation of the way they live and view the world. The texts were conceived primarily as spiritual exercises exercised by the author himself and then offered to disciples as a way of growing spiritually. Their value was often psycho-gogic, in the sense of leading the soul 'to school'.

A Socratic dialogue, for instance, when rightly interpreted in this context, is seen as a spiritual exercise practised in common, inciting readers to give attention to themselves, to take care of themselves, and to know themselves. Philosophers do not only know how to speak and debate but also how to live. They have the art of living. The ancients considered disciplines like ethics, logic and physics sub-disciplines of philosophy, all intended to form part of this art of living. For instance, they viewed physics not only as a way of producing theories about the nature of reality but also as a lived experience, a spiritual exercise which involved a way of seeing the world. It involved a cosmic consciousness that produces pleasure and joy for the soul.[2] Hadot finds justification for this line of interpretation in the very style of writing. The intention of the ancient author of making philosophy relevant for life, of making it a spiritual exercise rather than a dry system of propositions, is expressed in the dialogue form. A Socratic dialogue involves an interlocutor who prevents the dialogue from stalling on what is only theoretical and dogmatic. Some rhetorical slides from the theoretical exposition indicate that the real point is not the development of a discourse but a living of a life. Hadot appreciates ancient philosophy precisely because of this aspect:

> The concern with individual destiny and spiritual progress, the intransigent assertion of moral requirements, the call for meditation, the invitation to seek this inner peace, that all the schools, even those of the sceptics, propose as the aim of philosophy, the feeling for the seriousness and grandeur of existence – this seems to me to be what has never been surpassed in ancient philosophy and what always remains alive.[3]

1 Pierre Hadot, *Philosophy as a Way of Life: Spiritual Exercises from Socrates to Foucault*, ed. A.I. Davidson, trans. M. Chase, ed. A.I. Davidson (Oxford: Blackwell, 1995). See also: Pierre Hadot, *Exercices spirituels et philosophie antique* (Paris: Études Augustiniennes, 1987).

2 See P. Hadot, 'La philosophie antique: une éthique ou une pratique?', in Paul Demont (ed.), *Problèmes de la morale antique* (Amiens: Faculté des Lettres, 1993).

3 Hadot, *Philosophy as a Way of Life*, p. 69.

There are clear indications that many ancient philosophical traditions drew a clear distinction between discourse *about* philosophy and the *practice* of philosophy itself. Hadot's research shows, for instance, that the Stoics took theoretical philosophy as essentially a set of distinguishable sub-disciplines, each involving its own kind of explanation. Philosophy itself was not to be identified with these sub-disciplines. It was the organising accomplishment, in Hadot's words, 'a unique act, renewed at each instant'.[4] When we are engaged at this level of philosophy, we are not primarily concerned with theoretical logic, with the canons of correct reasoning, but with not letting ourselves be deceived in our everyday life by false representations or invalid inferences. At this level, we are no longer directly concerned with theoretical physics, with the various theories of the origin and evolution of the cosmos, but rather with being aware at every instant that we are *parts* of the world, and that we do better if we conform to this situation, or at least take it into consideration. When we are engaged at this level of philosophy, we no longer do ethical theory, in the sense of defining and classifying good actions and duties. We instead *act* in an ethical way.

Hadot expresses the novelty of this insight into the nature of philosophy by comparing ancient philosophical attitudes with modern ones. 'Ancient philosophy proposed to mankind an art of living. By contrast, modern philosophy appears above all as the construction of a technical jargon reserved for specialists.'[5] The various ancient philosophical schools can in fact be distinguished from each other by referring to their way of encouraging the neophyte to grow in wisdom. Socratism, Platonism, Aristotelianism, Epicureanism, Stoicism, and Scepticism, are models of life, fundamental forms in accordance with which reason may be applied to human existence. Hadot calls them archetypes of the quest for wisdom. For instance, Socratism concentrates on the dialogue as an exercise that makes the interlocutor put himself or herself in question so as to make the soul as beautiful and as wise as possible. For Plato, philosophy is a kind of training for death. For Aristotle, it is a life according to the mind. The Epicureans fixed their attention on the pleasure of existing, while the Stoics on the purity of intention. The Sceptics became famous for their struggle to achieve *ataraxia*, peace of the mind. We cannot assume, of course, that the list is closed. There is no reason to hold that in our day and age we are limited to merely rediscovering, or re-enacting, one or other of these models. It is a question of considerable importance to see whether the scientific revolution has given birth to a new 'wisdom-school'. This is precisely where the preceding inquiry on science and virtue finds considerable relevance.

Before engaging directly with this question, however, it is valuable to discuss what some commentators have seen as a drawback in Hadot's approach. Should the care of the self be considered exclusively a matter of the mind? Richard Shusterman, for instance, has accused Hadot of putting a one-sided emphasis on the mind.[6]

---

4 Ibid. p. 192.

5 Ibid. p. 272.

6 Richard Shusterman, 'Somaesthetics and Care of the Self: the case of Foucault', *The Monist*, 83/4 (2000): 530–51.

According to him, Hadot's approach ignores the body and defines the philosophical life wholly and emphatically in terms of intellectual expertise. Hadot seems to draw his basic inspiration primarily from the Stoics. This seems to limit his approach to a medical-therapeutic model of philosophical life dealing exclusively with the mind. This model, however, is not the only one possible. Michel Foucault has in fact interpreted the basic idea of care of the self in ancient authors not in the sense of healing or beautifying one's soul, but in the sense of aesthetics of the body. For Foucault, a very important theme of ancient philosophy is bodily pleasure.[7] He shows how, very early in Greek culture, the idea of the cultivation of the soul becomes impregnated with medical metaphors dealing with the care of the body. Philosophy and medicine are concerned with the same field. He recalls how philosophers used to employ various medical terms to describe the betterment of the soul, such as putting the scalpel to the wound, evacuating the superfluities, and so on. On the one hand, philosophers like Epictetus were keen on making students care for their bodies as much as for their souls. On the other hand, the physician Galen saw himself called to cure not just physical ailments but also the aberrations of the mind.[8] Considering the care of the self as a mental process seems therefore very limited.

This is the core of Shusterman's objection to Hadot's approach. On closer inspection, we can see that it is not as damaging as Shusterman thinks. What we have in this debate, in fact, is one possible reading of ancient trends presented against another. We are not obliged to assume that one reading is right and the other wrong. Hadot in fact confronts some points related to this objection in chapter seven of his book, a chapter that he dedicates to evaluating the contribution of Michel Foucault. Here, Hadot expresses his worry that a more bodily interpretation of the ancients' idea of care of the self would inevitably lead to a neglect of reason. He claims that 'by focussing his interpretation too exclusively on the culture of the self, the care of the self, the conversion toward the self – more generally by defining his ethical model as an aesthetic of existence – M. Foucault is propounding a culture of the self which is *too* aesthetic.'[9] The basic idea of both Hadot and Foucault is to tap into some uncharted sources of ancient literature not just for the fun of it but for the benefit of people living in the present day. Given that modern culture is definitely marked by the role of reason and experiment to an extent undreamed of by the ancient Greeks and Romans, at least some criteria for choice between the two interpretations can be discernible. If Foucault's interpretation is placed in the context of our day and age, the idea of 'care of the self' will be seen as too distant from the basic concerns of ordinary life. Hadot argues that a more authentic way of bringing ancient philosophical attitudes to bear on modern ways of living is to rediscover, with their help, the role of the philosopher as the one who is a practitioner of the exercise of wisdom. The philosopher aims at living concretely

7 M. Foucault, *The History of Sexuality, vol. 3, Care of the Self*, trans. R. Hurley (London: Allen Lane, 1988).

8 Ibid. pp. 55–56.

9 Hadot, *Philosophy as a Way of Life*, p. 211.

according to reason. This living certainly includes the management of one's body. The emphasis, however, is on *ascesis*, in the sense of exercise or training, rather than on *aesthetics*. Hadot's interesting suggestion is to accomplish this application of ancient attitudes to modern times by recalling the three exercises of Marcus Aurelius in his *Meditations*: first, the exercise of judging objectively, in accordance with inner reason; second, the exercise of acting in accordance with the reason which all human beings have in common; and third, the exercise of accepting the destiny imposed upon us by cosmic reason.[10] Whatever opinion one may have about the feasibility of this particular suggestion, one may safely conclude that Hadot's overall approach can certainly resist Shusterman. It's basic justification derives from the central role that reason and experiment enjoy in today's culture.

Science and Life

The main question to be confronted at this stage of my overall argument concerns the effects of natural science on the art of living. Can one speak of novelty with respect to what has already been excavated from ancient Greek philosophy as regards their various 'wisdom-schools'? Before answering this question, an initial problem has to be addressed. The very idea of investigating whether natural science can be a way of life, or whether it could incorporate a school of wisdom, seems to be self-contradictory because science, broadly speaking, has to do with knowledge of how things are, and knowledge of how things are does not correspond to what we normally mean by wisdom. Hinting that natural science might constitute a way of life, which I'm here taking to be practically the same thing as saying that it might constitute a school of wisdom, seems to be a non-starter. It is imperative therefore to engage in a preliminary investigation on the relation between knowledge and wisdom and then proceed to see how the discussion in previous chapters on virtue is helpful in this context.

This objection, in a way, has already been partially dealt with in Chapter 1. In that context, the problem was expressed in terms of the object of natural science, in the sense of what scientists are seeking. Scientific knowledge is geared towards the discovery of laws describing regularities. It is therefore geared towards the formulation of the most general principles of explanation. Contrary to this, the wise person is concerned with how to act in a particular situation. The reply offered in Chapter 1 consisted in recalling what a disposition is. Dispositions or habits are features of the person that are affected by what the person does. The effect repeated acts have on the person is observable only in the long run. The obtaining of scientific knowledge is one set of acts among others. Therefore, it is very plausible to hold that the dispositions of the person will be affected by 'scientific acts' just as by other

10 These points are evident, for instance, in Marcus Aurelius, *Meditations*, trans. A.S.L. Farquharson (Oxford: Oxford University Press, 1998), § 8,7. See also Pierre Hadot, 'La physique comme exercice spirituel: ou pessimisme et optimisme chez Marc Aurèle', *Revue de Théologie et de Philosophie*, 22 (1972): 225–39.

kinds of acts. One cannot hold that acts involved in attaining scientific knowledge are irrelevant to the dispositions of the person. This is the gist of the reply offered in Chapter 1. This reply can here be consolidated further by clarifying what we mean by the terms knowledge and wisdom. In the course of history, knowledge has often been described in terms of belief, for instance in the well-known formula 'true, justified belief'. Even apart from the Gettier problem, this definition does not exhaust all that we can say about knowledge. It is an atomistic definition that concentrates our attention on the particular utterance or proposition believed by a person. It relegates any consideration of the person himself or herself to the level of secondary importance or to the level of no importance at all. As a reaction to this, a recent recovery of the Aristotelian doctrine of intellectual virtues has made possible a much richer understanding of knowledge – knowledge as a *state of the person* who is engaged cognitively with the world in a way that goes beyond the state of mere believing.[11] In this sense, knowledge can be considered a state of a person in cognitive contact with reality in a manner that is good, desirable, and conducive to the attaining of genuine human flourishing. I am not interested here in arguing explicitly for one definition of knowledge and against the other. Such an argumentation, although useful for epistemologists, would take me too far from the line of reasoning being followed here. It is enough to rtain the point that a sustained analysis of the nature of knowledge obliges us to accept a significant link between knowledge and the person who knows.

If this link is manifest in the area of knowledge, it is even more marked in the area of wisdom. As has already been remarked, various authors in the course of history have emphasised the character of wisdom as *ars vivendi*, whereby intellectual activity such as Plato's contemplation of eternal, immutable Ideas, or the Aristotelian knowledge of first principles, is made relevant for everyday life. A typical statement highlighting this point is that of Cicero in his *Tusculan Disputations*:

> Philosophy dealt with numbers and movements, with the problem whence all things came, or whither they returned, and zealously inquired into the size of the stars, the spaces that divided them, their courses and all celestial phenomena; Socrates on the other hand was the first philosopher to call philosophy down from the heavens and set her in the cities of men and bring her also into their homes and compel her to ask questions about life, morality and things good and evil.[12]

The Renaissance development of this idea gave rise to the Prometheus-idea of the wise person who dares seize the torch of wisdom form the heavens to bring it down to poor mortals. What people must *know* to be wise is intimately related to what they must *do* to be wise. The intellectual and practical are considered mutually

11 For instance Linda Zagzebski, *Virtues of the Mind* (Cambridge: Cambridge University Press, 1996); Lorraine Code, *Epistemic Responsibility* (Hanover, N.H.: University Press of New England, 1987).

12 Marcus Tullius Cicero, *Tusculan Disputations*, trans. J.E. King (Cambridge Mass.: Harvard University Press, 1971), Book V, iv, §§ 10–11.

dependent.[13] While knowledge is often considered an entity having a cumulative nature resulting from a group effort, wisdom is a personal feature of the individual, somewhat like friendship. There is no stack of wisdom. There is no library of wisdom corresponding to our libraries of knowledge. Each person needs to attain wisdom as an individual achievement, perhaps helped by direct example from others or by suggestions embedded in narrative. Knowledge, in the analytic sense of true justified belief, is learnt piecemeal, often in the form of distinct propositions. Wisdom is a unifying feature of the person, bringing together the various elements of one's knowledge and the various experiences of one's life. To better clarify the difference between knowledge and wisdom, therefore, one should take the latter to be a habit of life that unites a reflective attitude with practical concerns. With it, individuals have the skill to judge their possibilities and thus determine a reasonable conception of the good life. Wisdom enables the individual evaluate complex situations of life in a way that attainment of the good life becomes realisable.

One crucial consequence is that science, knowledge and wisdom do not sit in separate compartments. Given that the knower is the very same person as the agent, and also the very same person who persists through time, an interaction between science, knowledge and wisdom is certainly possible and also very significant. Some relatively recent trends in science-studies have faced the challenge of seeing how natural science fares as a wisdom-school in various ways. For instance, some have tried to unmask harmful tendencies in the scientific mentality and call for a more humanitarian and critical science.[14] Others have emphasised that it is not enough to apply the scientific method to obtain knowledge. One must apply it also to build a better world in which people could flourish.[15] The major modes of engaging in this discussion seem to be two. One can either call for a re-evaluation of the aims of scientific inquiry. Or one can study the possible links between science and value. It is helpful to examine an example of each mode as an illustration.

The first example is Nicholas Maxwell. In his book *From Knowledge to Wisdom*, he argues that the burden of suffering and injustice carried by present masses of people is so extensive that humanity as a whole must give priority to the solving of the problems of life.[16] Inquiry is in need of a radical transformation in its aims. Instead of remaining geared towards obtaining knowledge, it needs to be channelled towards the enhancement of wisdom. He opposes what he calls the philosophy of knowledge and argues for the philosophy of wisdom. According to the former, the proper aim of rational inquiry is to acquire knowledge, and this can only be achieved

13 For a valuable study of the historical development of the idea of wisdom; see Eugene F. Rice, *The Renaissance Idea of Wisdom* (Cambridge Mass.: Harvard University Press, 1958).

14 See Jerry Ravetz, *Scientific Knowledge and its Social Problems* (Oxford: Clarendon, 1971).

15 For instance, Karl Popper, *The Open Society and its Enemies* (London: Routledge, 1995).

16 N. Maxwell, *From Knowledge to Wisdom: A Revolution in the Aims and Methods of Science* (Oxford: Basil Blackwell, 1984).

by dissociating the inquiry from distant goals, values and background beliefs related to common social life. He calls this dissociation an irrational feature of the philosophy of knowledge, and is convinced that it dominates most of present academic research and activity. Rationality requires that we articulate the basic problems facing us and then propose and assess possible solutions. When breaking up basic problems into smaller ones, we need to keep every aspect in view, so as not to be carried away by any one set of aspects and neglect the others. Rational inquiry demands keeping the correct balance. Unbalance very often results when organised inquiry is restricted to solving problems of knowledge in a way that is intellectually dissociated from problems of living.

Maxwell presents the philosophy of wisdom as a corrective to this. It enables the inquirer overcome the damaging effects of the philosophy of knowledge. The best way to describe this philosophy of wisdom is probably to refer to his concept of aim-oriented rationality. All problem-solving is aim-pursuing, but not all aim-pursuing is problem-solving. One may pursue an aim that is completely irrelevant to one's pressing problems. A young man may, for instance, be alienated by the immediate satisfaction available while playing games. He does not realise that his time may be better used to acquire a useful skill for future stable employment. Maxwell applies this point to the search for truth. Seeking truth for its own sake can always, and indeed should always, be supplemented by the meta-question: why are we pursuing this aim? This meta-question guarantees that our efforts as inquirers are channelled towards the obtaining of valuable truths, valuable for genuine human flourishing. Neglecting this meta-question deforms rational inquiry not only in science but also in philosophy. Maxwell insists that philosophy has sadly been contaminated beyond recognition by a limited idea of science: 'All of life, and all of inquiry, has suffered to a greater or lesser extent as a result of the intellectual failure of philosophy to give priority to the task of promoting wisdom in life' (p. 149).

Some may take exception here by arguing that relevant knowledge must precede action for that action to be rational. Research in natural science is value-neutral and a clear distinction needs to be drawn between the work of scientists and that of politicians who decide how to use the fruits of scientific research. Maxwell blocks this objection in a compelling way. He recalls how, in basic everyday practice, we are always discrediting this principle. We do not depend on correct prior knowledge of the internal structure of the eye to see well. We do not depend on correct prior knowledge of the structure of the brain to think well. Similarly, we do not depend on correct, prior knowledge of the world to act. Knowledge is useful, certainly. Nevertheless, it cannot be considered prior to action at the level of rudimentary everyday practice. In more sophisticated action, the kind that constitutes scientific practice, the theory-ladenness of observation demands that a simplistic model involving a dichotomy between science and life be abandoned. Maxwell summarises his argument this way: 'What really matters is what we do, what goes on in our lives. Knowledge is of importance insofar as it contributes to, and participates in, life'(p. 181).

The basic idea therefore is related to the general theme of this chapter, namely the link between science and life. My proposal was to consider this link via the idea of virtue, while Maxwell is exploring it via an evaluation of the aims of inquiry. It is not difficult to see how his aim-oriented rationality is in fact a good habit. He is advocating a revision of our habits, by proposing an awareness of what is a real virtue and what isn't. He explicitly states this, albeit rather rapidly, at one stage in his argument: 'The vital point is to promote in society the *habit* of putting forward and criticising proposals for action intended to help achieve what is of value [italics in the original]' (p. 197). Given that society has reality only in as far as it is a group of individuals, each with his or her own personality, the conclusion to be drawn here is that Maxwell's approach is *contained* in the virtue approach, rather than the other way round. This is not to say that Maxwell's analysis is superfluous. It is important as an attempt to articulate in detail what some of the virtues related to the scientific mentality are. It throws light, moreover, on the significant point that science, if engaged in with the right frame of mind, can be considered a wisdom-school on a par with those of Ancient Greece. It remains true, however, that the main focus of such a discussion should not be abstract discourse but the individual as a person living through time.[17]

I will turn now to another contemporary author. This is to illustrate the other way of linking science with genuine human flourishing. Hugh Lacey, in his book on the possible connections between science and value, concentrates on the question of whether science is value-free.[18] The first task he undertakes is the one of giving an acceptable account of values and, more specifically, of cognitive values. In this latter category he puts the characteristics that make beliefs or theories 'good', in the sense of rationally acceptable. So he effectively takes a value to be a property of the object confronting the human subject. The value is cognitive when the object is a belief or theory. He analyses the question of science and value by refracting it into three possible modes. One can understand science as value-free in the sense of it being impartial, or in the sense of it being neutral, or in the sense of it being autonomous. Impartiality refers to the state where values are not among the grounds for accepting and rejecting theories. Neutrality, roughly speaking, refers to consistency with all value judgements. Autonomy refers to the fact that the scientific community conducts its investigations in self-governed institutions free from outside interference. Lacey's book is an attempt at elucidating these modes of understanding science as value free. His conclusions show that while autonomy cannot be sustained the other two aspects can.

---

17 Another difference between my approach and Maxwell's, a difference I'm not dwelling on for lack of space, has to do with the overall approach. My main interest is to explore how the scientific mentality impinges, positively or negatively, on the character traits of the person. His interest, on the contrary, is to campaign for a revolution *within* the scientific mentality.

18 Hugh Lacey, *Is Science Value Free? Values and Scientific Understanding* (London: Routledge, 1999).

A central idea that runs right through this discussion is that of materialist strategies. Lacey uses this expression to highlight the fact that the descriptive language in which the data for science are expressed contains what he calls materialist terms: quantitative and mathematical terms applicable in virtue of measurement and experimental operations. Modern scientific practice shows considerable variety as regards description, explanation, structure, or law. Nevertheless, there is a common feature corresponding to these materialist strategies that indicate the kinds of data acceptable for scientific theories. Only theories resulting from materialist strategies have cognitive value. This central idea is not new. Lacey's original contribution lies in his claim that such materialist strategies are just one kind among several other kinds of strategy that can be adopted without sacrificing the basic requirement of intersubjectivity of empirical data. His major argument goes this way. Materialist strategies are adopted within the scientific community for a reason. They are adopted because they reinforce modern values of control that are evident in the desire to 'grasp' the world. Understanding here is taken to be a matter of control. A problem arises because the world understood in this way is not constituted of objects that are purely and simply given, ontologically independent of human observers. Objects are always handled, manipulated, measured or experimented upon via operations involving groups of people. The materialist strategies give the impression of doing away completely with all subject-related terms. This, however, is only an illusion. If Lacey is right, the world is made up of objects that are partly constituted in practices, and these practices have mutually reinforcing interactions with the modern values of control: we understand the world by choosing, consciously or unconsciously, what we want to grasp and how. This does not mean that the door is closed for genuine discovery. It means that anyone making new discoveries cannot affirm or conclude that the underlying strategies were unique and independent of human subjectivity.

Lacey proceeds by making a number of suggestions as regards alternative strategies. The most interesting of these seems to be the 'grassroots empowerment approach'. This strategy is not presented as a radical substitute for the materialist strategies mentioned above. It is meant to be a meta-strategy to which the materialist strategies are seen as subordinate. The basic idea here is that, in general, the aim of gaining understanding provides by itself no direction to scientific inquiry. In order to pursue it, one needs to choose and follow a particular approach to inquiry. An approach consists in strategies that limit the kinds of theories that can be entertained. A strategy is what gives direction to research. The objective of the grassroots empowerment approach to science involves identifying the object of research in line with potential value for local well-being and community. An example would be the aim of identifying objects of research relevant to agricultural practices. The overall approach is not exclusively linked to the control of nature. One aims to control nature in view of higher values. It is this higher level of strategic planning that shows that science cannot be considered autonomous. It can be considered neutral and impartial, in the senses explained above. It cannot however be considered value-free in the sense of conducting its investigations in self-governed institutions free from outside interference.

The basic thesis of Lacey's work, therefore, is certainly related to the general theme of this chapter, namely the link between science and life. His emphasis is on value, while my approach highlights the link between science and life via the idea of virtue. He explores the issue of strategies, as they are embedded in practices expressing various values. I concentrate on the individual person, the individual agent within such practices. There is some overlap between the two approaches as regards the general aim of working towards genuine human flourishing. The question of value, however, in the way Lacey discusses it, is not centred on the person but on the *object* of the action of the person. As I pointed out above, he takes a value to be a property of the object confronting the human subject. His analysis therefore is certainly valuable but does not substitute the analysis engaged in so far concerning how the scientific mentality affects the individual person.[19]

To recapitulate, therefore, the line of argument in this section has been as follows. The aim at the beginning was to show what it means to hold that science, knowledge and wisdom do not sit in separate compartments. In the previous section, I gave priority to the fact that the person as knower is the same person as agent with the same identity persisting through time. Two major ways of highlighting this basic point in current literature are the re-evaluation of the aims of scientific inquiry, and the discovery of the links between science and values. The former is well represented by Maxwell's approach, the latter by Lacey's. I proceeded then by giving a quick overview of these approaches so as to illustrate that, although valid and useful, they are incomplete. They cannot substitute the virtues approach developed in the previous chapters. The three are complementary approaches: the first is methodological in character, the second is a meta-scientific in the sense of seeking the criteria of the use of such methods; the third is a personalist approach, concentrating on the individual.

What are, therefore, the effects of natural science on the art of living, when this art is taken as a personal quest? Can one speak of novelty with respect to what had already been excavated from ancient Greek wisdom-schools? I mentioned above how the ancient philosophical schools can be distinguished from each other by referring to their way of encouraging the novice to grow in wisdom. Historians identify such schools as models of life, archetypes of the quest for wisdom.

From my investigation so far, we can highlight the five major virtues associated with natural science as follows. First, the virtue of living with due respect both towards common sense, what I called the manifest image, and to the more sophisticated scientific world-views, the scientific image. Second, the virtue of living in a way that gives due importance both to what is universal and to what is particular. Third, the virtue of prudential risk-taking that is aware of both the strong

19 The only point where Lacey makes an explicit reference to the individual person is when he recalls that the activities and virtues involved in the gaining of scientific knowledge are considered by some philosophers, such as Hilary Putnam, as being partly constitutive of human flourishing (p. 105). See also: Hilary Putnam, *Reason, Truth and History* (Cambridge: Cambridge University Press, 1981), p. 134.

points and weak points of our intellectual faculties. Fourth, the virtue of living in a way that acknowledges the context-dependence of all explanation, the scientific included. Fifth, the virtue of heuristic courage in the course of history. Is there a way of condensing these various aspects into one characteristic? The best candidate seems to be the last virtue in the list: heuristic courage. This, in a way, includes the others. It is reasonable to conclude that the basic novelty brought about by the rise of natural science in the realm of the art of living has to do with the dimension of heuristic courage, as it is expressed in a practical interaction with the material world. As historians have shown in various ways, the need for heuristic courage has been highlighted in the course of history in proportion to the growth of distance between the manifest and scientific images of the world. As we advance further and further towards the extremely large and towards the extremely small, as we let science take us further and further away from our everyday environment, the more heuristically courageous we need to become.

Science, the Good Life, and God

Because of the personalist dimension of the approach in this book, interesting new horizons open up as regards the interaction between science and theology. It is worthwhile at the very start to fix the basic framework for discussion by clarifying the term 'theology'. Although a vast discipline, it can be described in broad strokes as discourse on God. The classic definitions emphasise the aspect of faith seeking understanding, *fides quaerens intellectum*. In this sense, theology is the business of people who have faith in the first place, and who want to see how what they believe and what they live makes sense, or is in line with their other beliefs. For contemporary people, these 'other beliefs' can be considered a set of beliefs constituted mainly of the results of natural science. Theology in the classic sense of *fides quaerens intellectum*, therefore, has to confront the on-going challenge of figuring out ways of understanding the faith so as to help the believer avoid a fragmentation of his or her personal set of beliefs – to avoid a state of intellectual schizophrenia. One important point to note here is that the concept of faith, as used here, is much broader than that of a set of propositions. Believers have a specific way of life that manifests itself in various ways. Only one of these manifestations is in the form of linguistic expression. Moreover, only part of this linguistic expression is formulated in definite propositional form that constitute a creed. Faith is often better seen as a form of life rather than as a set of beliefs.

It is precisely because faith is a form of life that the entire issue of virtue becomes crucial. This point is again related to the major strategy of seeing the individual's life as a whole. Hence I do not concentrate exclusively on one particular scientific procedure or act, but on the long effects the scientific mentality may have on the life of a person. I do not concentrate exclusively on one particular religious act but on the long term effects faith may have on the life of a person. The reason for doing a particular act is that we attach a value to something we consider the end in relation to that act. Seeing the scientist's life as a whole, rather than considering merely a time-slice of it, means acknowledging that, over and above subsidiary ends there is an ultimate end. In fact, I have endeavoured to classify the character-traits resulting from the scientific mentality as virtues or vices in terms of their being in line or not in line with a person's ultimate end.

Even though there is no clear consensus among philosophers as regards what constitutes the ultimate end, Aristotle's arguments in the *Nicomachean Ethics* are still compelling. He adopts what has been called a synoptic ethical approach.[20] He conceives philosophy as a complete system linking ethical and scientific understanding so as to determine what constitutes a worth-while human life. He recalls that the ultimate end is happiness, and tries to derive the nature of happiness from what is specifically human, namely reason. 'If happiness is activity in accordance with virtue, it is reasonable that it should be in accordance with the highest virtue; and this will be that of the best thing in us.'[21] The perfection of reason, the contemplation of the truth, is therefore what constitutes happiness for humans. Moreover, if we follow Aristotle in this line of reasoning, we realise that what is ultimately desirable is a certain kind of happiness that is, first, permanent, in the sense that it cannot be lost once acquired, and, second, continuous, as opposed to episodic. Less than this will not do. Aristotle sees here a conflict between, on the one hand, the ideal happiness our reasoning demands and, on the other hand, what can be expected within a human life, limited as it is. Human happiness will always be an imperfect realisation of the ultimate end. His arguments lead to the conclusion that our concept of happiness exceeds our practical possibilities. He writes: 'Such a life will be too high for man; for it is not in so far as he is man that he will live so, but in so far as something divine is present in him.'[22] Medieval philosopher-theologians like St Thomas Aquinas have seen in this claim an important way of harmonising the Aristotelian natural way of understanding human good with the Christian version of ultimate end as attainable through the grace of God. The philosophical approach prepares the mind to accept the theological account. The ultimate end, instead of being seen as contemplation of the truth can be seen as the person's meeting with God, the *summum bonum*.

A minimal conclusion can be safely drawn in the following terms. If it is plausible to hold that natural science is relevant to the personal character of the individual scientist in so far as it can hinder or promote genuine human flourishing, then it is also plausible to hold that natural science is relevant for the individual's search for God, understood as the ultimate end, the *summum bonum*.

Is this claim original in the area of science and theology? A good way how to appreciate the element of originality of this personalist-ethical way of dialogue between the two disciplines is by situating it within the landscape of the various

20 The term 'synoptic ethics' is used by John Cottingham, *Philosophy and the Good Life* (Cambridge: Cambridge University Press, 1998). See especially chapters 1 and 3.

21 Aristotle, *Nicomachean Ethics*, 1177[a], 12–13.

22 Ibid., 1177[b], 26; see also: T. Nagel, 'Aristotle on Eudaimonia', in A.O. Rorty (ed.), *Essays on Aristotle's Ethics* (Berkeley: University of California Press, 1980), pp. 1–6.

research strategies that have been inaugurated up to now. Many scholars working in this area spend considerable intellectual energy in trying to forge a naturalistic theology consonant with the result of current natural science. A good example is the work by W.B. Drees.[23] According to him, theology needs first to avoid the strong version of naturalism that makes *all* legitimate questions scientific ones. It then needs to endorse a weak version of naturalism whereby non-material aspects of reality, like music, are not considered meaningless but considered 'embodied' within elements forming part of the natural sciences domain. This is typical of this approach. Hence for music, say, we have ink on paper, or vibrations of strings. As a consequence, the usual working hypothesis for this approach is that the domain of science underpins all other domains. One discusses time, for instance, by bringing together the concept of time in the New Testament and the latest insights from modern physics, with the hope that the understanding of the former will be enriched by the understanding of the latter. This approach exemplifies the interaction between science and theology on the conceptual level. Valuable interaction can occur, it is said, because of the fact that the two disciplines share some common concepts. The theological one is said to be embodied within the scientific one. Although clear as a research programme, such an approach needs careful handling because the word 'embodied' is misleading. Whatever science will tell us about the nature of the ink, or about the elasticity of the material of the string, cannot help us much as regards appreciating the music. Likewise, what science tells us about the nature of time cannot help us much as regards appreciating the biblical meaning of time. Careful analysis of the relation between thing and meaning is therefore essential in this approach.

Another possible approach, still not well developed, is the one highlighting the interaction between science and theology not as regards their claims, but as regards their questions.[24] Instead of discussing the reality or meaning of entities, one may discuss the reality of questions for the scientific or the theological community. A range of questions are real in a given community at a given time when they are questions which the members of the community can see how they can get to grips with them. Understanding questions real in my community implies that I appreciate what the community considers relevant to those questions. Understanding questions unreal in my community but real in another community involves appreciation of the considerations that would be taken in that other community to be relevant to that question. This approach makes both scientists and theologians aware of higher criteria determining how their research should develop, and where their research resources should be invested. The major challenge in this approach is how to avoid a

23 Willem B. Drees, *Religion, Science and Naturalism* (Cambridge: Cambridge University Press, 1996).

24 I explore this with particular reference to the environment issue in L. Caruana, 'Questions concerning Science, Theology, and the Environment', *Gregorianum*, 79 (1998): 149–61. I draw inspiration from the original discussion as regards scientific questions in N. Jardine, *The Scenes of Inquiry* (Cambridge: Cambridge University Press, 1991).

radical conventionalism that may creep in if the reality of entities is made to depend on that of questions.

The consideration of science and virtue constitutes a third, distinct way of working in the area of science and theology. Both science and faith are here considered not merely as sets of claims but as agglomerations of practices linking language to ways of behaviour. The focus of attention is on the individual. Character-traits are affected both by the scientific mentality and by other belief-domains that the individual inherits from life in general, the religious belief-set included. If both these effects are seen as determining the basic desire to become a good person, a new door is open for further investigation. Both science and theology can be considered frameworks for a mentality that affects the way we live.

Appendix A

The Hippocratic Oath

Although this oath has been considered the model of medical conduct for many centuries, its author and its year of origin remain unknown. It is divided into eight parts of which the first is the oath proper, and the last a kind of prayer or invocation. The promise is made around two main subjects: conduct towards teachers and students, and conduct towards patients and their households. Ludwig Edelstein's translation is the following:

> I swear by Apollo Physician and Asclepius and Hygieia and Panaceia and all the gods and goddesses, making them my witnesses, that I will fulfil according to my ability and judgement this oath and this covenant:
>
> To hold him who has taught me this art as equal to my parents and to live my life in partnership with him, and if he is in need of money to give him a share of mine, and to regard his offspring as equal to my brothers in male lineage and to teach them this art – if they desire to learn it – without fee and covenant; to give a share of precepts and oral instruction and all the other learning to my sons and to the sons of him who has instructed me and to pupils who have signed the covenant and have taken an oath according to the medical law, but no one else.
>
> I will apply dietetic measures for the benefit of the sick according to my ability and judgement; I will keep them from harm and injustice.
>
> I will neither give a deadly drug to anybody who asked for it, nor will I make a suggestion to this effect. Similarly I will not give to a woman an abortive remedy. In purity and holiness I will guard my life and my art.
>
> I will not use the knife, not even on sufferers from stone, but will withdraw in favour of such men as are engaged in this work.
>
> Whatever houses I may visit, I will come for the benefit of the sick, remaining free of all intentional injustice, of all mischief and in particular of sexual relations with both female and male persons, be they free or slave.
>
> What I may see or hear in the course of the treatment or even outside of the treatment in regard to the life of men, which on no account one must spread abroad, I will keep to myself, holding such things shameful to be spoken about.
>
> If I fulfil this oath and do not violate it, may it be granted to me to enjoy life and art, being honoured with fame among all men for all time to come; if I transgress it and swear falsely, may the opposite of all this be my lot.[1]

---

1 Ludwig Edelstein, *The Hippocratic Oath: Text, Translation, and Interpretation* (Baltimore: Johns Hopkins Press, 1943). For this appendix, I am also indebted to Leon R.

A very significant point I would like to highlight for the purpose of my overall inquiry is that the oath represents an excellent example of the personal involvement within the practice of a science. The oath, in fact, does not indicate where the technical ends and where the ethical begins. All the requirements flow equally from the basic idea that medicine is not merely an ethically neutral technique but an activity established and empowered by a notion of the good. Consider, for instance, the injunction not to give a woman an abortive medicine. The issue here is less related to the act of abortion itself, its moral or legal dimension, and more to the integrity of the doctor. The Oath is highlighting the incompatibility between two factors. On the one hand, we have medicine as a practice constituted by the assistance to living nature in humans in view of maintenance, and perpetuation. On the other hand, we have the destruction of nascent life. There is outright inconsistency here. The spirit of the Oath includes not only the promise to refrain from participating in such practices as abortion, which taint the purity of the physician as an individual, but also the desire to inform and edify the entire community as regards the inestimable value of life. The final invocation introduces the time dimension. The one taking the Oath contemplates how things will probably turn out in the long run, and expresses various intentions in this regard. In this aspect as well, the idea of medical practice is here presented as related to the life of virtue. The fundamental point being referred to is that the good person, physician or not, is fully recognisable as such only at the end of his or her life.

Kass, *Toward a more Natural Science: Biology and Human Affairs* (New York: Free Press, 1985), chapter 9. For a fuller treatment of the points mentioned here, see also: J. Chadwick and W.N. Mann (eds), *The Medical Works of Hippocrates* (Oxford: Blackwell, 1950); Fabrice Jotterand, 'The Hippocratic Oath and Contemporary Medicine: Dialectic between Past Ideals and Present Reality?' *Journal of Medicine and Philosophy*, 30 (2005): 107–28.

Appendix B

Elements of a Logic of Virtue

In what follows, the expression 'logic of virtue' is meant to refer to the kind of thinking that occurs in agents as they strive to live virtuously. The treatment will often refer to the classic Aristotelian insights in the *Nicomachean Ethics*, but is not primarily intended as an analysis of Aristotle's theory of the mean. Whether Aristotle would recognise the following proposal as his own, or as a legitimate continuation of his own, will not be my main concern. Although consonance with Aristotle's theory would certainly be added support to this proposal, I will not be seeking such support directly. The proposal is meant to speak for itself. Its justification is meant to lie principally in its internal coherence and in its usefulness as a fertile model for further inquiry.

A good way to start is to identify two areas of philosophy that offer interesting insights into the logic of virtue: the *logica negationis* as used in Negative Theology, and the more general *reductio ad absurdum*. I will examine briefly each one in turn. In philosophy of religion, the method called *via negationis* consists in denying of God anything that belongs to a contingent being as such. In this way, we do not show what God is but what He is not. This differs from the normal way of inquiry. The normal way is to start from a definition and then add positive differences so as to gain better and better understanding. Given that the essence of God is inaccessible, however, this strategy cannot be used for the case of God. The only way forward is to collect a series of negative differences that show what this essence of God is not. The resulting knowledge of God is not positive. It is a string of negatives: God is known as immutable, as incorporeal, as immaterial, and so on. In spite of this, however, by denying all the limitations found in creatures, the *via negationis* allows philosophers of religion to say with ever greater precision what God is not and what God cannot possibly be. Hence, by distinguishing God from what is not God, they attain some knowledge of what God is. We have here therefore a classic example of getting some conceptual grasp of an idea by working in from the extremes.

This strategy in philosophy of religion is analogous to what is going on in our thinking when we consider a virtue as the mean between two extremes. Courage is considered as the mean disposition between two extreme dispositions one of rashness, the other of cowardice. The point I want to highlight here is that the account of the virtue flows, as it were, from the extremes inwards. It starts from an idea of the vices, which are understandable extremes, and ends up with an idea of virtue. Before embarking on the task of making this more precise, let us draw some more inspiration from the other area of philosophy mentioned above: *reductio ad absurdum*.

As a mode of argumentation, the *reductio ad absurdum*, which is found repeatedly in Aristotle's *Prior Analytics*, can have various forms depending on how the consequence is shown to be untenable. In the strictest construction of a *reductio* argument, the consequence is untenable because it is a self-contradiction. In less strict constructions, the consequence is untenable because it is a falsehood (*reductio ad falsum* or *ad impossibile*) or an implausibility (*reductio ad ridiculum* or *ad incommodum*). It is useful, therefore, to distinguish between at least three kinds of *reductio* arguments: those depending on logical analysis, those depending on empirical observation, and those depending on practical considerations.

As regards the first kind, the one involving logical analysis and eventual self-contradiction, the classic example occurs in Greek mathematics. Pythagoras discovered the incommensurablity of the diagonal of a square with its sides by first supposing that the diagonal is indeed commensurable and then showing that this leads to a logical contradiction. As regards the second kind of *reductio* arguments, those depending on empirical observation, W. Kneale and M. Kneale make some useful remarks. They explain how, in metaphysics, the first precise meaning of the term 'dialectic' seems to be '*reductio ad impossibile*'.[1] This is distinguishable from *reductio ad absurdum* in so far as it involves drawing consequences from the hypothesis that is simply false, and not necessarily self-contradictory, as in the second. For instance, in Plato's *Meno* 93, Socrates argues that, if virtue were teachable, good men would instruct their sons in it. It is, however, a well-known fact, Socrates continues, that Pericles and others did not succeed in making their sons virtuous. Here, an empirically observable consequence is drawn from the premise 'Virtue is teachable'. Moreover, this empirical consequence is known to be false, in the sense that it is contradicted by an empirical fact, and hence the premise is refuted. There is no logical inconsistency as such in holding that 'Virtue is teachable'. The problem arises when observation is brought into the argument. Had Socrates never heard of Pericles and his sons' lack of virtue, it would not have been possible for him to argue along these lines. This kind of argument, therefore, depends on the reliability of the knowledge claims shared by the protagonists of the debate.

The third kind of *reductio* argument depends on practical considerations. This kind is often neglected. Since most uses of the *reductio* style of argument are related to logic or mathematical reasoning, the full range of application is rarely acknowledged. It is essential, however, to acknowledge that a genuine *reductio* argument may be extended to cover instructions, practices and policies. In such cases, a specific *modus operandi* is shown to result in absurdity. Consider, for instance, a maxim of the following kind: 'Never say never'. This is evidently self-defeating in practice. In less evident cases, the absurdity or anomaly of the procedure becomes manifest only when the actual adoption of the procedure and its implementation are analysed carefully. Immanuel Kant's doctrine of the categorical imperative is essentially a kind of *reductio* argument applied to modes of procedure. For instance, the practice

1 W. Kneale and M. Kneale, *The Development of Logic* (Oxford: Clarendon Press, 1962), p. 9.

of lying involves a rule, namely 'Tell lies when it suits your convenience', that will lead to an absurdity when generalised. The absurdity in this case is equivalent to the annihilation of verbal communication. Kant's famous insight lies here. His proposal can be described in terms of *reductio* reasoning with practical overtones. He is effectively saying that rule-conforming practices that reduce, on generalisation, to an absurd or impossible situation show themselves morally unacceptable.[2]

This third kind of *reductio* argument offers a valuable starting point for formulating the logic of virtue. The other insight gained previously from my brief consideration of the *logica negationis* involved the strategy of working in from the extremes. Both these insights constitute the foundations of the technique I'm using in this book. This technique involves what may be called *reductio ad impedimentum*: a kind of *reductio ad impossibile* related to practical considerations. It is a kind of *reductio* that is dependent both on our theoretical knowledge of the world, and also on the practical possibilities it allows.

The objective I had set myself at the beginning involved an attempt at formalising the logic of virtue. The idea is to gain some insights into the structure of the kind of thinking that occurs in agents as they strive to live virtuously. This is now possible to some extent. Consider the simpler cases first. What is happening when I want to act courageously? In a typical situation open to the possibility of acting courageously, I come to realise I am not at all courageous but indeed a coward when I look back, as it were, and become aware that I could have been less fearful, or more passionate. In other words, I realise I am acting like a coward when I look back and see that I am getting too far from acting like a rash person. If we take the two extreme habits of cowardice and rashness as two opposing poles, we can say that the logic of virtue, at least in this simple example, involves the realisation of being too close to one pole and too far away from the opposite pole. The reasoning here involves the imagination. This faculty enables me to reply to crucial questions related to counterfactual situations involving the two poles, namely: (1) what actions would eventually make me cowardly here? (2) what actions would eventually make me rash here? The faculty of imagination enables me to have some picture of these situations. The ensuing strategy would then involve avoiding both extremes.

This project, as it is described up to now, is of course different from Aristotle's. He was engaged, primarily, on the task of understanding what we mean by virtue. His aim seems to have been to arrive at an ethical theory that explains well the various virtues we identify in normal circumstances. He takes our coherent talk about virtue as a sign that there is an essence of virtue to be discovered. My project, as opposed to this, is a kind of formalisation that will allow, I hope, a clarification of the process of thinking about specific instances of action in line, or out of line,

2 For more insights into the logic of *reductio* arguments, see M. Dorolle, 'La valeur des conclusions par l'absurde', *Révue philosophique*, 86 (1918): 309–13; J.M. Lee, 'The Form of a reductio ad absurdum', *Notre Dame Journal of Formal Logic*, 14 (1973): 381–86; G. Ryle, 'Philosophical Arguments', in *Colloquium Papers*, vol. 2 (Bristol: University of Bristol, 1992), pp. 194–211.

with virtue. Aristotle's focus was on the nature of virtue. Mine is on the way we think about it and manifest it in action, not in the abstract, but in particular cases. I must, to be sure, be careful not to try to render into mathematical form what is unmathemisable. A full algorithm that would allow the identification of virtuous action in any given situation is a pure fiction. The formalisation I'm after is valuable only to the modest extent that it demonstrates how the imagination is at work in fixing the mean by referring to two opposing extreme situations.

Although my main intention, as mentioned at the very start, is not to enter into the debate about what Aristotle had in mind, it may be useful to note here that, according to J.O. Urmson, my approach deviates from Aristotle's in another way. Urmson proposes that 'it must be wrong to hold that Aristotle thought that we should decide how to act on particular occasions by working in from the extremes'.[3] Urmson's argument is based on the distinction between what he calls a disposition towards the mean and a mean disposition. Excellence of character is a mean disposition from among many possible dispositions. It is not a disposition towards mean or intermediate emotions. It is not, therefore, the same as the doctrine of moderation, according to which the agent in any particular situation should act in line with the maxim: 'neither too much, nor too little'. I find Urmson's argument interesting but too blunt to capture all the important nuances. He does not take into consideration what Aristotle would include in the exercise of practical wisdom for a particular occasion. A particular exercise of wisdom includes skills like planning ability, experience, executive ability, and so forth. These skills often involve the consideration of possible future situations where some variables are changed in the imagination while others are kept constant. This exercise of these skills, therefore, is effectively a way of 'working in from the extremes'. Moreover, Urmson does not give enough importance to the fact that agents have no control over the development of their dispositions if not via the control they have over singular actions. The principle of moderation is therefore always made use of by the wise person.

Let me investigate further, therefore, the role of intelligence and deliberation in the exercise of practical wisdom. Suppose, in a most simple situation, that a measure can be defined along a two dimensional space corresponding to a particular passion, P. Passion is here intended in the Aristotelian sense: we can have too much of it or too little. We may be tempted to let the strength of the passion range from 0 to 1. This would imply that the virtue would correspond to a situation where $P = 0.5$. This temptation should be resisted because it distorts what we want to describe. In a real life situation, we have certainty neither as regards the vices nor as regards the virtue. In other words, in a given situation, say in a rescue operation, there is no clear fixed *point* of reference that enables the person to see what would constitute a

3 J.O., Urmson, 'Aristotle's Doctrine of the Mean', in A.O. Rorty (ed.), *Essays in Aristotle's Ethics* (Berkeley: University of California Press, 1980), p. 162. For a critique of this article see R. Hursthouse, 'A False Doctrine of the Mean', *Proceedings of the Aristotelian Society*, 81 (1980–1): 57–72.

cowardly response, or a rash one, or a courageous one. What we have are degrees of possibilities merging into one another.

The best model therefore seems to be the determination of the mid-point of a line *AB* given that we are limited to two-dimensional space starting at some unknown point *C* on the line. This will involve an analogy derived from navigation. In fact, Aristotle, faced with the inherent imprecision involved in his ethical inquiry, suggests precisely such a model of navigation. He writes: 'matters concerned with conduct and questions of what is good for us have no fixity, any more than matters of health. The general account being of this nature, the account of particular cases is yet more lacking in exactness; for they do not fall under any art or set of precepts, but the agents themselves must in each case consider what is appropriate to the occasion, as happens also in the art of medicine or of navigation.'[4] So consider a man that is on a very long straight road. He knows that the emergency phone is found in a small box at the middle of the road. His car breaks down somewhere along the road, but he cannot really say *where* he's situated with respect to the phone. So he proceeds this way. He observes that on the side of the long road there are electricity poles at equal distances from each other. He uses his binoculars to view the two ends of the road *A* and *B*. If the poles look smaller at *A* than at *B*, he walks towards *A*. If the poles look smaller at *B* than at *A*, he walks towards *B*. He checks again and again until the distant poles at *A* look the same size as those at *B*. This reassures him that he has arrived at the mid-point of *AB*.

Now apply this model to the case of virtue. Using the binoculars corresponds to imagination. This is effectively situating oneself in a counterfactual situation by extrapolating the particular passion that drives my action. Consider the case discussed in Chapter 2. I act on the belief that the scientific image has a revisionary importance of degree *D*. To arrive at the virtuous balance, I exercise my faculty of imagination. I imagine what it would be like had I acted on the belief that the scientific image has a revisionary importance of degree *D* as *D* tends to infinity. In the same way, I can also imagine what it would be like as *D* tends to zero. Each counterfactual situation is evaluated, via the imagination, by seeing how it affects my general aim of flourishing well as a human being in the long run. The correct evaluation of the counterfactual situation depends on various factors. The most significant of these is the knowledge I have of the causal nexus that determines the run of things in various situations. A lack of such knowledge would undermine the value of my faculty of imagination. It would make it impossible for me to determine how things would be like in a counterfactual situation. And this would undermine therefore my capacity to determine where virtue lies. To some extent therefore this proposal of a logic of virtue does include some important elements of the Socratic idea of virtue as

4 *Nicomechean Ethics*, 1104^{a}, 3–1104^{a}, 9. Aristotle probably derived his model of navigation from Plato, *Republic* IV and VI. See also W.A. Welton and R. Polansky, 'The Viability of Virtue in the Mean', *Aperion*, 28 (1995): 79–102; reprinted in L.P. Gerson (ed.), *Aristotle: Critical Assessments*, vol. 3 (London, New York: Routledge 1999), pp. 292–311.

knowledge. If my conclusion doesn't allow me to hold that virtue is identical to knowledge, it shows nevertheless that they are intimately related.

Up to now I have employed a model involving continuous change of parameters. For a formalisation, however, a discrete analysis is better. Hence there are four steps to be made.

1. Assume that life-situations can be characterised as possible worlds, with names p_i, in which the person acts, or imagines he or she would be in a position to act, with a certain amount of passion or emotion. Using Urmson's terminology, we can say that the kind of emotion determines the field of a particular virtue.[5] We are simplifying matters here by assuming that there is only one major emotion per possible world. In real life, several different continua of variable passions may come into play in regard to a single situation. For instance, a rescue operation often involves more kinds of passions than just the one we associate with courage.

2. Assume also that these possible worlds can be conceived as discrete situations ranging from $p_{-\infty}$ to $p_{+\infty}$.

3. Define a logical predicate of degree one, $\mathcal{A}$, that needs the logical name of a situation (or possible world) to form a sentence. We may call this the *antiaretic predicate*. $\mathcal{A}p$ means '*p* hinders the attaining of genuine human flourishing'.

4. For simplicity, assume we are dealing with ideal agents whose imagination has all the resources needed to determine, with a reasonable degree of plausibility, the truth-value of the sentences formed by the predicate $\mathcal{A}$ and each of the possible worlds; in other words, assume that, for a given situation p_i, it is always possible to say whether or not $\mathcal{A}p_i$.

From here, we can think of life-situations as two opposing sequences departing from a point:

$$\ldots p_{-3}, p_{-2}, p_{-1}, p_0, p_1, p_2, p_3, \ldots$$

The initial point, p_0, corresponds to where the protagonist starts the inquiry, the actual world.

5 Urmson, 'Aristotle's Doctrine of the Mean', p. 163.

The idea of finding the mean between two extremes corresponds, therefore, to the following:

1. $\exists\, k$ such that $\forall\, i \geq k$, $\mathcal{A}p_i$
2. $\exists\, r$ such that $\forall\, j \leq r$, $\mathcal{A}p_j$
3. hence, choose life-situation p_s, where $|k - s| = |r - s|$

Notice that there is no assumption that where one starts in this reflection, i.e. p_0, is the life-situation of a virtuous person. The fact that s may not be zero shows that the person may learn, or grow in virtue, through this process.

Bibliography

Aristotle, *The Basic Works of Aristotle*, ed. R. McKeon (New York: Random House, 1941).

Audi, R., 'Acting from Virtue', *Mind*, 104 (1995): 449–71.

———, *Epistemology* (London: Routledge, 1998).

Aurelius, M., *Meditations*, trans. A.S.L. Farquharson (Oxford: Oxford University Press, 1998).

Bacon, F., *Novum Organum*, trans. P. Urbach and J. Gibson (La Salle, IL.: Open Court, 1994).

Batterman, R., 'Multiple realizability and universality', *British Journal for the Philosophy of Science*, 51 (2000): 115–45.

———, *The Devil in the Details* (Oxford: Oxford University Press, 2002).

Bennett, M.R. and Hacker, P.M.S., *Philosophical Foundations of Neuroscience* (Oxford: Blackwell, 2003).

Bernstein, R., *Beyond Objectivism and Relativism* (Oxford: Basil Blackwell, 1983).

Boyd, R., 'Scientific Realism and Naturalistic Epistemology', in P.D. Asquith and T. Nickles (eds), *PSA 1980*, vol. 2 (East Lansing, MI: Philosophy of Science Association, 1981), pp. 613–62.

———, 'Realism, Approximate Truth and Philosophical Method', in C.W. Savage (ed.), *Scientific Theories* (Minneapolis: University of Minnesota Press, 1990), pp. 355–91.

Brenner, A.A., 'Holism a Century Ago: The Elaboration of Duhem's Thesis', *Synthese*, 83 (1990): 325–35.

Britz, C., and Brink, C., 'Computing verisimilitude', *Notre Dame Journal of Formal Logic*, 36 (1995): 30–43.

Bunzl, M., *The Context of Explanation* (Dordrecht, London: Kluwer Academic Publishers, 1993).

Carr, D., *Educating the Virtues: An Essay on the Philosophical Psychology of Moral Development and Education* (London: Routledge, 1990).

Caruana, L., 'Questions concerning Science, Theology and the Environment', *Gregorianum*, 79 (1998): 149–61.

———, 'Is Science eliminating ordinary talk?' *Forum Philosophicum*, 4 (1999): 25–39.

———, *Holism and the Understanding of Science* (Aldershot: Ashgate, 2000).

Chadwick, J. and Mann, W.N. (eds), *The Medical Works of Hippocrates* (Oxford: Blackwell, 1950).

Cicero, M.T., *Tusculan Disputations*, trans. J.E. King (Cambridge Mass.: Harvard University Press, 1971).

Code, L., *Epistemic Responsibility* (Hanover, N.H.: University Press of New England, 1987).

Cohen, L.J., *The Implications of Induction* (London: Methuen, 1970).

Cottingham, J., *Philosophy and the Good Life* (Cambridge: Cambridge University Press, 1998).

Craig, E.J., 'The Problem of Necessary Truth' in: S. Blackburn (ed.), *Meaning, Reference and Necessity* (Cambridge: Cambridge University Press, 1975), pp. 1–31.

Davidson, D., 'On the very Idea of a Conceptual Scheme', *Proceedings and Addresses of the American Philosophical Association*, 57 (1974): 5–20; reprinted in: D. Davidson, *Inquiries into Truth and Interpretation* (Oxford: Clarendon Press, 1984), pp. 183–98.

Dent, N.J.H., *The Moral Psychology of the Virtues* (Cambridge: Cambridge University Press, 1984).

Dorolle, M., 'La valeur des conclusions par l'absurde', *Révue philosophique*, 86 (1918): 309–13.

Drees, W.B., *Religion, Science and Naturalism* (Cambridge: Cambridge University Press, 1996).

Dretske, F., *Seeing and Knowing* (London: Routledge and Kegan Paul, 1969).

Dupré, J., *The Disorder of Things: Metaphysical Foundations of the Disunity of Science* (Cambridge, Mass.; London: Harvard University Press, 1993).

Eddington, A., *The Nature of the Physical World* (Cambridge: Cambridge University Press, 1928).

Edelstein, L., *The Hippocratic Oath: Text, Translation, and Interpretation* (Baltimore: Johns Hopkins Press, 1943).

Edwards, P., 'Russell's Doubts about Induction', *Mind*, 68 (1949): 141–63.

Feigl, H., 'De principiis non disputandum...? On the Meaning and the Limits of Justification', in M. Black (ed.), *Philosophical Analysis: A Collection of Essays* (Ithaca, NY: Cornell University Press, 1950), pp. 119–56.

Feyerabend, P., *Against Method* (London: Verso, 1993).

Fine, A., 'Piecemeal Realism', *Philosophical Studies*, 61 (1991): 79–96.

Fodor, J. 'Special Sciences, or the Disunity of Science as a Working Hypothesis', *Synthese*, 28 (1974): 97–115.

Foucault, M., *The History of Sexuality, vol. 3, Care of the Self*, trans. R. Hurley (London: Allen Lane, 1988).

Fraassen, B. van, *The Scientific Image* (Oxford: Oxford University Press, 1980).

———, *The Empirical Stance* (New Haven: Yale University Press, 2002).

Geach, P., *The Virtues* (Cambridge: Cambridge University Press, 1977).

Goodman, N., *Facts, Fiction and Forecast* (London: Athlone Press, 1954).

Hacking, I., *Representing and Intervening* (Cambridge: Cambridge University Press, 1983).

Hadot, P., 'La physique comme exercice spirituel: ou pessimisme et optimisme chez Marc Aurèle', *Revue de Théologie et de Philosophie*, 22 (1972): 225–39.

———, *Exercices spirituels et philosophie antique* (Paris: Études Augustiniennes, 1987).

———, 'La philosophie antique: une éthique ou une pratique?', in Paul Demont (ed.), *Problèmes de la morale antique* (Amiens: Faculté des Lettres, 1993).

———, *Philosophy as a Way of Life: Spiritual Exercises from Socrates to Foucault*, trans. M. Chase, ed. A. I. Davidson (Oxford: Blackwell, 1995).

Hanson, N.R., *Patterns of Discovery: an Inquiry into the Conceptual Foundations of Science* (Cambridge: Cambridge University Press, 1958).

Harrison, R., *On what there must be* (Oxford: Clarendon Press, 1974).

Hodge, M.J.S., and Cantor, G.N., 'The Development of Philosophy of Science since 1900', in R.C. Olby, G.N. Cantor, J.R.R. Christie and M.J.S. Hodge (eds), *A Companion to the History of Modern Science* (London: Routledge, 1996), pp. 838–52.

Holmes, F.L., *Lavoisier and the Chemistry of life: an Exploration of Scientific Creativity* (Wisconson: University of Wisconsin Press, 1985).

Hoyningen-Huene, P., *Reconstructing Scientific Revolutions* (Chicago: Chicago University Press, 1993).

Hume, D., *A Treatise of Human Nature* (London: Fontana, 1970–1972).

———, *An Enquiry concerning Human Understanding: A Critical Edition*, ed. Tom L. Beauchamp (Oxford: Clarendon, 2000).

Hunt, L., 'Character and Thought', *American Philosophical Quarterly*, 15 (1978): 177–86.

Hursthouse, R., 'A False Doctrine of the Mean', *Proceedings of the Aristotelian Society*, 81 (1980–1): 57–72.

———, 'Virtue Theory and Abortion', *Philosophy and Pubic Affairs*, 20 (1991): 223–46.

———, *On Virtue Ethics* (Oxford University Press, 1999).

Jardine, N., *The Scenes of Inquiry* (Cambridge: Cambridge University Press, 1991).

Jotterand, F., 'The Hippocratic Oath and Contemporary Medicine: Dialectic between Past Ideals and Present Reality?' *Journal of Medicine and Philosophy*, 30 (2005). 107–28.

Kant I., *Prolegomena to any Future Metaphysics that will be able to present itself as a Science*, trans. P.G. Lucas (Manchester: Manchester University Press, 1953).

Kass, L.R., *Toward a more Natural Science: Biology and Human Affairs* (New York: Free Press, 1985).

Kenny, A., 'The Homunculus Fallacy', in *The Legacy of Wittgenstein* (Oxford: Blackwell, 1989), pp. 125–36.

Kim, J., 'Multiple Realization and the Metaphysics of Reduction', *Philosophy and Phenomenological Research*, 52 (1992): 1–26.

Kitcher, P. and Salmon, W., 'Van Fraassen on Explanation', *Journal of Philosophy* 84 (1987): 315–30.

Kneale, W. and Kneale, M., *The Development of Logic* (Oxford: Clarendon Press, 1962).

Kuhn, T.S., *The Structure of Scientific Revolutions*, 2nd ed. enl. (Chicago: University of Chicago Press, 1970).

———, 'Objectivity, Value Judgement and Theory Choice', in *The Essential Tension: Selected Studies in Scientific Tradition can Change* (Chicago: University of Chicago Press, 1977).

Lacey, H., *Is Science Value Free? Values and Scientific Understanding* (London: Routledge, 1999).

Lakatos, I., 'Falsification and Methodology of Scientific Research Programmes,' in I. Lakatos and A. Musgrave (eds), *Criticism and Growth of Knowledge* (Cambridge: Cambridge University Press, 1970).

———, 'History of Science and its Rational Reconstructions' in R. Buck and R. Cohen (eds), *Boston Studies in the Philosophy of Science*, vol. 8 (Dordrecht: D. Reidel, 1971).

———, *The Methodology of Scientific Research Programmes*, ed. J. Worrall and G. Currie (Cambridge: Cambridge University Press, 1978).

Laudan, L., 'A confutation of convergent realism', in J. Leplin (ed.), *Scientific Realism* (Berkeley: University of California Press, 1984).

Lee, J.M., 'The Form of a reductio ad absurdum', *Notre Dame Journal of Formal Logic*, 14 (1973): 381–86.

Lloyd, E.A. and Anderson, C.G., 'Empiricism, Objectivity, and Explanation', *Midwest Studies in Philosophy*, 18 (1993): 121–31.

Lombard, P., *Sententiae in IV libris distinctae, ad fidem codicum antiquiorum restituta* (Grottaferrata: Ed. Collegii S. Bonaventurae ad Charas Aquas, 1971–81).

MacIntyre, A., *After Virtue: a Study in Moral Theory* (Notre Dame, Indiana: University of Notre Dame Press, 1981).

Maxwell, N., *From Knowledge to Wisdom: A Revolution in the Aims and Methods of Science* (Oxford: Basil Blackwell, 1984).

McDowell, J., 'Virtue and Reason', *The Monist*, 62 (1979): 331–50, reprinted in R. Crisp and M. Slote (eds), *Virtue Ethics* (Oxford: Oxford University Press, 1998), pp. 141–62.

Meerbote, R., 'Kant's Use of the Notions "Objective Reality" and "Objective Validity"', *Kant-Studien*, 63 (1972): 51–58.

Meilaender, G., *The Theory and Practice of Virtue* (Notre Dame Indiana: University of Notre Dame Press, 1984).

Melsen, A.G. van, *Science and Responsibility* (Pittsburgh: Duquesne University Press, 1970).

Mill, J.S., *A System of Logic, Ratiocinative and Inductive: being a connected View of the Principles of Evidence and the Methods of Scientific Investigation*, ed. J. M. Robson (Toronto: University of Toronto Press, 1978).

Murdoch, I., *The Sovereignty of Good* (London: Ark, 1970).

Nagel, T., 'Aristotle on Eudaimonia', in A.O. Rorty (ed.), *Essays on Aristotle's Ethics* (Berkeley: University of California Press, 1980), pp. 1–6.

Niiniluoto, I., 'Verisimilitude: the third period', *British Journal for the Philosophy of Science*, 49 (1998): 1–29.

O'Hear, A., *Karl Popper* (London: Routledge, 1980).

Peirce, C.S., 'Pragmaticism and Critical Common-Sensism' in C. Hartshorn, P. Weiss and A. Burks (eds), *Collected Papers*, vol. 5 (Cambridge, MA: Harvard University Press, 1931–1958), pp. 346–50.

———, 'Consequences of Critical Common-Sensism' in: C. Hartshorn, P. Weiss and A. Burks (eds), *Collected Papers*, vol. 5 (Cambridge, MA: Harvard University Press, 1931–1958), pp. 351–75.

Plato, *The Republic*, trans. G.M.A. Grube, in J.M. Cooper (ed.), *Plato: Complete Works* (Indianapolis: Hackett Publishing Com. Ltd., 1977), pp. 971–1224.

Polanyi, M., *Personal Knowledge* (London: Routledge and Kegan Paul, 1958).

Popper, K., *Conjectures and Refutations: the Growth of Scientific Knowledge*, 3rd ed. revised and enlarged (London: Routledge and Kegan Paul, 1969).

———, 'Moral Responsibility of the Scientist', in P. Weingartner and G. Zecha (eds), *Induction, Physics and Ethics* (Dordrecth: Reidel, 1970).

———, 'Replies to my Critics' in P.A. Schilpp (ed.), *The Philosophy of Karl Popper* (Illinois: Open Court, 1974), pp. 961–1197.

———, *The Logic of Scientific Discovery* (London: Routledge, 1992).

———, *The Open Society and its Enemies* (London: Routledge, 1995).

Psillos, S., *Scientific Realism: How Science tracks the Truth* (London: Routledge, 1999).

Putnam, H., *Meaning and the Moral Sciences* (London: Routledge & Kegan Paul, 1978).

———, *Reason, Truth and History* (Cambridge: Cambridge University Press, 1981).

Railton, P., 'Probability, Explanation, Information', *Synthese*, 48 (1981): 233–56.

Ravetz, J., *Scientific Knowledge and its Social Problems* (Oxford: Clarendon, 1971).

Rice, E.F., *The Renaissance Idea of Wisdom* (Cambridge Mass.: Harvard University Press, 1958).

Rock, I., *Perception* (New York: Scientific American Books, 1984).

Rohrlich, F., 'Pluralistic Ontology and Theory Reduction in the Physical Sciences', *British Journal for the Philosophy of Science*, 39 (1988): 295–312.

Rose, D., 'A Portrait of the Brain', in R. Gregory, J. Harris, P. Heard, and D. Rose (eds), *The Artful Eye* (Oxford: Oxford University Press), pp. 28–51.

Ruben, D.-H., *Explaining Explanation* (London: Routledge, 1992).

——— (ed.), *Explanation* (Oxford: Oxford University Press, 1994).

Ruse, M. and Wilson, E.O., 'Moral Philosophy as Applied Science', *Philosophy*, 61 (1986): 173–92.

Ryle, G., 'Philosophical Arguments', in *Colloquium Papers*, vol. 2 (Bristol: University of Bristol, 1992), pp. 194–211.

Salmon, W., 'Four decades of Scientific Explanation' in P. Kitcher and W. Salmon (eds), *Scientific Explanation* (Minneapolis: University of Minnesota Press, 1989), pp. 3–196.

Sanders, A.F., *Michael Polanyi's Post-Critical Epistemology* (Amsterdam: Rodopi, 1988).

Scriven, M., 'Explanation, Prediction and Laws', in H. Feigl and G. Maxwell (eds), *Scientific Explanation, Space and Time* (Minneapolis: University of Minnesota Press, 1962), pp. 170–230.

Sellars, W., *Science, Perception, and Reality* (London: Routledge & Kegan Paul, 1968).

Shusterman, R., 'Somaesthetics and Care of the Self: the case of Foucault', *The Monist*, 83/4 (2000): 530–51.

Skyrms, B., *Choice and Chance: an Introduction to Inductive Logic* (Belmont USA: Wadsworth, 4th ed., 2000).

Smith, A.D., 'Of Primary and Secondary Qualities', *Philosophical Review*, 99 (1990): 221–55.

Söderqvist, T., 'Virtue-Ethics and the Historiography of Science', *Danish Yearbook of Philosophy*, 32 (1997): 45–64.

Sorell, T., *Scientism* (London: Routledge, 1991).

Stenmark, M., *Scientism: Science, Ethics and Religion* (Aldershot: Ashgate, 2001).

Stroud, B., *The Quest for Reality* (Oxford: Oxford University Press, 2000).

Thomas Aquinas, *Summa Theologiae*, Latin text and English translation, introductions, notes, appendices and glossaries (London: Blackfriars, 1964–1981).

Tichý, P., 'On Popper's definition of Verisimilitude', *British Journal for the Philosophy of Science*, 25 (1974): 155–60.

Urmson, J.O., 'Some Questions concerning Validity', in A. Flew (ed.), *Essays in Conceptual Analysis* (London: MacMillan, 1956), pp. 120–33.

———, 'Aristotle's Doctrine of the Mean', in A.O. Rorty (ed.), *Essays in Aristotle's Ethics* (Berkeley: University of California Press, 1980), pp. 157–70.

Uttal, W.R., 'Mind, the psychobiology of', in G. Adelman and B.H. Smith (eds), *Encyclopedia of Neuroscience*, vol. II (Amsterdam: Elsevier 1999), pp. 1185–7.

Welton, W.A. and Polansky, R., 'The Viability of Virtue in the Mean', *Aperion*, 28 (1995): 79–102; reprinted in L.P. Gerson (ed.), *Aristotle: Critical Assessments*, vol. 3 (London, New York: Routledge 1999), pp. 292–311.

Wittgenstein, L., *Philosophical Investigations*, trans. G.E.M. Anscombe (Oxford: Blackwell, 1953).

Wright, G.H. von, *The Varieties of Goodness* (New York: Routledge & Kegan Paul, 1963).

Zagzebski, L.T., *Virtues of the Mind* (Cambridge, Cambridge University Press, 1996).

———, 'The Search for the Source of Epistemic Good', *Metaphilosophy*, 34 (2003): 12–28.

Index